TOOLS FOR INNOVATION

TOOLS FOR INNOVATION

Edited by

ARTHUR B. MARKMAN
KRISTIN L. WOOD

OXFORD
UNIVERSITY PRESS

2009

OXFORD
UNIVERSITY PRESS

Oxford University Press, Inc., publishes works that further
Oxford University's objective of excellence
in research, scholarship, and education.

Oxford New York
Auckland Cape Town Dar es Salaam Hong Kong Karachi
Kuala Lumpur Madrid Melbourne Mexico City Nairobi
New Delhi Shanghai Taipei Toronto

With offices in
Argentina Austria Brazil Chile Czech Republic France Greece
Guatemala Hungary Italy Japan Poland Portugal Singapore
South Korea Switzerland Thailand Turkey Ukraine Vietnam

Published by Oxford University Press, Inc.
198 Madison Avenue, New York, New York 10016

www.oup.com

Library of Congress Cataloging-in-Publication Data

Tools for innovation / edited by Arthur B. Markman, Kristin L. Wood.
p. cm.
Includes index.
ISBN 978-0-19-538163-4
1. Creative thinking. 2. Design, Industrial. 3. Creative ability. 4. Cognitive
science—Research. I. Markman, Arthur B. II. Wood, Kristin L.
BF408.T62 2009
153.3′5—dc22
2009001719

Printed in the United States of America
on acid-free paper

10 09 08 07 06 05 04 03

To Lucas, 'Eylam, and Niv: Innovators all

Contributor List

Srinivasan Anandan	Graduate Research Associate, Department of Mechanical Engineering, Clemson University, Clemson, South Carolina
Stuart R. Borrett, Ph.D.	Assistant Professor of Biology and Marine Biology, University of North Carolina, Wilmington, Wilmington, North Carolina
Will Bridewell, Ph.D.	Postdoctoral Research Scientist, Computational Learning Laboratory, Center for the Study of Language and Information, Stanford University, Stanford, California
Bo T. Christensen, Ph.D.	Associate Professor, Copenhagen Business School, Copenhagen, Denmark
Darren W. Dahl, Ph.D.	Fred H. Siller Professor in Applied Marketing, Sauder School of Business, The University of British Columbia, Vancouver, British Columbia
Dan Jensen, Ph.D.	Professor of Engineering Mechanics, United States Air Force Academy, Colorado Springs, Colorado
Andruid Kerne, Ph.D.	Associate Professor, Department of Computer Science, Interface Ecology Lab, Texas A&M University, College Station, Texas
Eunyee Koh, Ph.D.	Research Scientist, Department of Computer Science, Interface Ecology Lab,

	Texas A&M University, College Station, Texas
Pat Langley, Ph.D.	Consulting Professor of Symbolic Systems, Head of Computational Learning Laboratory, Center for the Study of Language and Information, Stanford University, Stanford, California
Jeffrey P. Laux	Doctoral Student, Psychology, University of Texas at Austin, Austin, Texas
Julie S. Linsey, Ph.D.	Assistant Professor of Mechanical Engineering, Texas A&M University, College Station, Texas
Arthur B. Markman, Ph.D.	Annabel Irion Worsham Centennial Professor of Psychology and Marketing, University of Texas at Austin, Austin, Texas
C. Page Moreau, Ph.D., M.B.A.	Associate Professor of Marketing, Leeds School of Business, University of Colorado, Boulder, Colorado
Jeremy T. Murphy	Doctoral Student, Mechanical Engineering, University of Texas at Austin, Austin, Texas
Christian D. Schunn, Ph.D.	Associate Professor of Psychology, Learning Systems, and Intelligent Systems, University of Pittsburgh, Pittsburgh, Pennsylvania
Jami Shah, Ph.D.	Professor and Director, Design Automation Laboratory, Arizona State University, Tempe, Arizona
Vikramjit Singh, M.S.	Doctoral Student, M.O.R.P.H. Lab, Manufacturing and Design Research Laboratory, Department of Mechanical Engineering, The University of Texas, Austin, Texas

Steven M. Smith, Ph.D.	Professor of Psychology, Interface Ecology Lab, Department of Psychology, Texas A&M University, College Station, Texas
Joshua D. Summers, Ph.D.	Associate Professor, Department of Mechanical Engineering, Clemson University, Clemson, South Carolina
Masaki Suwa, Ph.D.	Professor of Information and Intelligence, Chukyo University, Toyota, Aichi, Japan
Sudhakar Teegavarapu	Graduate Research Associate, Department of Mechanical Engineering, Clemson University, Clemson, South Carolina
Barbara Tversky, Ph.D.	Professor Emerita of Psychology, Stanford University, Professor of Psychology and Education, Columbia Teachers College, Columbia University, New York, New York
Brandon Walther	Doctoral Student, M.O.R.P.H. Lab, Manufacturing and Design Research Laboratory, Department of Mechanical Engineering, The University of Texas, Austin, Texas
Thomas B. Ward, Ph.D.	Professor of Psychology, University of Alabama, Tuscaloosa, Alabama
Robert Weisberg, Ph.D.	Professor of Psychology, Temple University, Philadelphia, Pennsylvania
Kristin L. Wood, Ph.D.	Cullen Trust Endowed Professor in Engineering and University Distinguished Teaching Professor of Mechanical Engineering at the University of Texas at Austin, Austin, Texas

Acknowledgments

This book came about as the result of a workshop called Tools for Innovation held at the University of Texas. We had been talking about finding a way to get psychologists, engineers, computer scientists, and consumer behavior researchers together to talk about innovation. John Sibley Butler and Robert Peterson of the IC^2 Institute at the University of Texas were intrigued by this idea, and they generously gave us money to host this conference. Both of them also gave generously of their time to help with conference organization. In addition, Coral Franke of the IC^2 Institute provided logistical support that made the conference a success. Finally, the National Science Foundation provided additional funding to help graduate students and young faculty attend the workshop.

Thanks to Erin Spalding for her help organizing the chapters and getting them ready for publication. Julie Linsey and Jeff Laux provided a lot of support for the conference. The whole Similarity and Cognition lab read the chapters and provided feedback that was passed along to the chapter authors. And of course, thanks to the authors as a group for providing such a great collection of chapters.

At Oxford, Catharine Carlin was very helpful in getting this project into the OUP fold. Abby Gross read over the manuscript and gave the authors valuable feedback. Mark O'Malley guided us through the production process.

Finally, Art Markman would like to acknowledge the support of the W.W. Heath Centennial Fellowship in the IC2 Institute, and Kris Wood would like to acknowledge the support of the Cullen Endowed Professorship in Engineering.

Contents

TOOLS FOR INNOVATION

CHAPTER 1

THE COGNITIVE SCIENCE OF INNOVATION TOOLS

ARTHUR B. MARKMAN
KRISTIN L. WOOD

"A PSYCHOLOGIST and an engineer sit down to write a paper." Rather than being the setup to a joke, this state of affairs reflects what we see as the fundamental mode of research for studying the process of innovation. In particular, innovation research lies at the nexus of basic cognitive science and content domains in which people are going to generate novel creative products. It is at this nexus where the field can go beyond merely elucidating the basic cognition underlying creativity, to generating proposals for tools that can support the creative process.

This book presents a collection of chapters that lie at the leading edge of research on innovation and tools to support innovation processes. Much of this work reflects collaborations between scientists with different types of expertise. For example, the chapter by Smith, Kerne, Koh, and Shah reflects a collaboration between people with expertise in psychology, engineering, and computer science. The chapter by Tversky and Suwa involves a collaboration between a psychologist and an information scientist. The work by Dahl and Moreau brings together two researchers in consumer behavior who have a strong background in psychology. These examples demonstrate the importance of thinking broadly about creativity in order to make headway on what may be the most daunting question in psychology: "Where do new ideas come from?" and one of the most daunting question in domain-specific design: "How do we develop methods and tools that enhance and empower designers to create novel ideas?"

In this chapter, we set the stage for the rest of the book by starting with a brief definition of creativity and innovation. A more elaborate discussion of innovation and creativity is presented in the chapter by Weisberg. Then we discuss the importance of interdisciplinary research for developing tools for innovation. This section certainly discusses advantages of interdisciplinary research for the study of innovation tools. However, this work also focuses on the importance of research on innovation for basic research in the component disciplines of cognitive science. Often, people involved in basic research view the flow of information from basic science to applied science without recognizing the role that applications of basic research can play in basic science.

Finally, we identify aspects of the innovation process that seem particularly promising for generating tools. Some of these areas are ones that are represented with chapters in this volume. Others are important avenues of future research.

Innovation and Creativity

Figure 1–1 shows a classic image of a lone individual endeavoring to create. This model is being called into question as cognitive science research illuminates our understanding of creative cognition. For one thing, creativity does not happen in isolation, but rather in groups. For another, creativity requires a lot of information about the domain being studied. Despite our greater understanding of creativity, however, an agreed-upon definition for innovation and creativity is elusive. There are probably as many definitions of creativity and innovation as there are researchers who study this process (and probably more). In the next chapter, Weisberg discusses definitions of innovation and creativity in detail. We will not commence that discussion here.

In practical settings, however, there are three crucial aspects of innovation that are important. First, for a solution to some kind of practical problem to be innovative, it must be truly novel in the history of the field. For example, in patent law, an invention or process can be patented only if it is not previously known to individuals skilled in the art. That is, if the idea

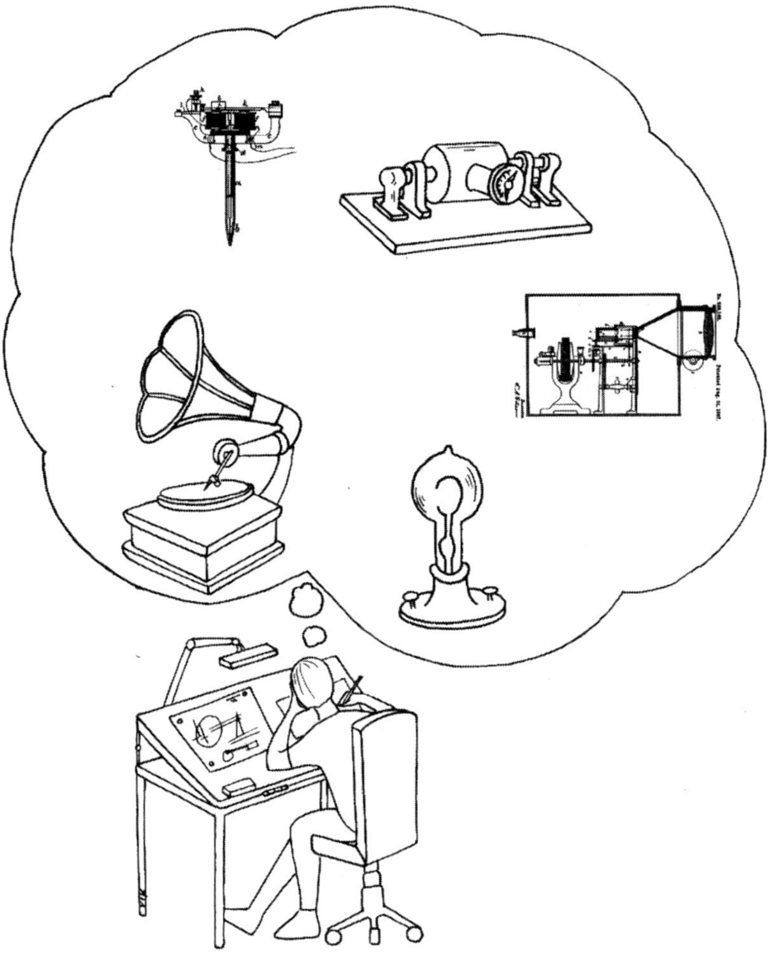

Figure 1–1 A snapshot of an individual engaged in the creative process.

already exists in the field, then it is not "creative" in the practical sense. Boden (1994) refers to this type of creativity as *historical* creativity. That is, it is the first instance of an idea's being generated by anyone. She contrasts historical creativity with *personal* creativity, in which an individual has an idea that is new for them.

The second critical aspect of innovation is that it must address the problem being approached. One way that innovation differs from creativity more generally is that creative acts may be undertaken without any particular goal in mind. A musician may have a goal for a new piece of music, but it is certainly possible to try to create a new piece of music with no particular aim in mind. In contrast, innovation typically begins with a statement of a problem to be solved.

It is always possible that a truly creative idea will emerge from work on a specific problem, though the idea itself does not solve the problem at hand. A classic example of this situation is in the development of Post-It® notes.[1] In 1968, Spencer Silver was looking for a formula for a very strong adhesive. In the process, one compound they tried tended to form little balls. The bond was not strong, but it did seem to be reusable. Because the project being solved at that time was to develop a strong adhesive, the reusable adhesive would not really qualify as an innovation, because it did not solve an existing problem. Silver did think the new compound was likely to be useful, so he made it known around the company that he had found a reusable adhesive. Sometime later, another 3M employee, Art Fry, was trying to find a reusable bookmark to keep his place while singing in a church choir. He solved this problem using Silver's reusable adhesive. This problem solution would qualify as an innovation by the definition we are using, because it provided a novel solution to a specific problem.

The third aspect of "innovation" is that the idea must be able to be successfully implemented. Because innovation happens in the context of a practical problem, innovations must do more than provide a theoretical solution to a problem. They must provide a solution that can be implemented. Often, this implementation must address many additional problems that go beyond the initial creative spark that led to the general idea for a solution. Indeed, in the story of the Post-It® note, while the idea itself was hatched in 1974, it took three more years before 3M was able to begin manufacturing the product, because a number of additional technical problems had to be addressed. All of these issues are central to real innovation.

The purpose of giving this three-part definition of innovation is simply to provide a framework for thinking about issues related to studying innovation and creativity and for developing tools to support the innovation process. We realize that not everyone will agree with the three principles in this section. Indeed, not all of the authors in this volume may agree. Nonetheless, this list provides a starting point for further discussion.

[1] The full story is available at the 3M website, www.3m.com. In addition, refer to Shaw, 2002.

Where Can New Ideas Come From?

In order to find ways to improve innovation processes, it is important to think about where new ideas can come from. Then tools can be developed that are tailored to these settings. Three potential sources of new ideas are serendipity, research and development, and reasoning.

Serendipity in innovation

One source of new ideas is serendipity. The example of 3M and the reusable adhesive described in the previous section has this element. A corporate research team was seeking a strong adhesive. In the course of working on this project, they stumbled on another compound with desirable properties, though not the desirable properties they were seeking at the time.

By its nature, of course, it is not possible to control serendipity. It is possible, however, to analyze prior cases of serendipitous findings to maximize the likelihood that future cases will lead to fruitful outcomes. In the 3M example, the discovery of the reusable adhesive was serendipitous, but the rest of the story required a lot of effort. Spencer Silver spent considerable time shopping the compound around the company. When Art Fry had a need the adhesive could be used for, the idea was known widely enough for him to be able to try it out.

More generally, the reusable adhesive was a side effect of the normal research process for 3M. A compound with properties that might be desirable for another use was discovered. Success for 3M required a scientist who took it upon himself to ensure that the compound was ultimately given a use. One potential avenue for innovation tools, then, would be to create effective methods for making potential solutions to problems available throughout an organization to maximize the value of serendipitous findings.

Research and development in innovation

A second source of innovative ideas emerges through the research and development process. When a problem has been identified, a systematic research process can be quite effective in producing a solution. For example, in Chapter 2, Weisberg discusses some of the work that the Wright brothers did in the development of the airplane. One difference between the Wrights and other teams that were working to develop heavier-than-air flying machines was the systematic way that they developed the various systems that were required to generate lift and control the flight of the plane. Through an understanding of the basic physical principles involved in the systems, the Wrights were able to develop testable insights that led to innovative outcomes.

Systematic research and development has now become the norm in corporate research. Within the research and development setting, however, there are a number of processes that could be made more efficient. Similar to the problem identified with serendipitous findings, it is important to create systems that large organizations can use to find the results of corporate research and development to maximize the likelihood that discoveries can be used to solve problems across the organization. Often, large organizations have a vertical structure in which advances within a particular research group are not well known outside of the group. Thus, the same tools that can publicize serendipitous findings can also publicize findings that are the expected outcome of research. Chapter 10 by Summers, Anandan, and Teegavarapu discusses tools that can be used to help members of an organization find existing designs that may help them solve a new design problem.

In addition, the research process is only as good as the framing of the problem that is used to generate the research. Thus, it is often important to refine the way problem statements are generated at the start of a research project. Chapter 2 by Weisberg discusses the issue of problem statements in some more detail. In this section, however, there are two further issues of importance.

First, problem statements often embed within them assumptions about the proper way to solve a problem. For example, consider the aphorism, "Build a better mousetrap, and the world will beat a path to your door." The statement to "build a better mousetrap" contains the assumption that the best solution the problem of ridding a house of mice is to trap the mice effectively. Of course, there are many solutions to keeping a house mouse-free. One potential solution is to prevent mice from even entering the house. A second is to create a way to induce mice that enter a house to leave. A third is to find a benefit of having mice in the house and reap the benefit of mouse-infestation. The point of this (somewhat fanciful) discussion is that individuals and teams engaged in a process of innovation should analyze their problem statements to uncover hidden assumptions that have become part of the framing of the innovation task. By so doing, they may remove or strategically attack biases and potential fixations.

A related issue is that problem statements are often too vague to be addressed effectively by research. This issue can be seen in the example of the Wright brothers. They did not set about trying just to create a single machine that could fly. Instead, they conceptualized the airplane as a set of intersecting systems. By breaking down the aircraft into a set of smaller subsystems, the Wright brothers were able to focus specific research on different aspects of the device, including the shape of the wings and the control structure of the plane. Systematic tests, based on fundamental physical insights, were developed for each aspect of the device.

In addition, problem statements may be vague, because they do not take into account the role of a device in a broader system. For example, Bradshaw (2007) has examined the work of Edison and his lab as they developed a viable incandescent light. Many groups had recognized that running a current through a filament caused the filament to shine. The key problem was that the filament would quickly burn out, making the light impractical. In the years before finding a workable design, Edison and his team worked with designs that had a regulator that would cut off the current when the filament got too hot. This solution led the light bulb to flicker.

Prior to finding a better design, however, Edison also began to develop a system for delivering electricity to homes that would allow them to have the power needed for these lights. Transporting power over large distances required higher voltages than most groups working on this problem were using. Thus, Edison further constrained the problem to be "to find an electric light that worked at relatively high voltages whose filament would not burn out." This change required increasing the resistance of the light bulb (so that it would work at high voltages), which in turn required a long filament. Having a long filament led to experiments with silicon insulation of the filament, which then led to the use of carbon filaments. Ultimately, a carbon filament was part of the successful design.

The key to this example is that most inventors who were trying to develop a viable electric light were focused on the light itself. Edison considered both the light and the distribution system, which put additional constraints on the form of the bulb. Ultimately, these additional constraints helped Edison and his team discover a solution. This example suggests that innovation teams engaged in research and development must consider both the particular device that is the focus of research and also the broader system in which the device is embedded. This broader system may introduce additional limitations that (paradoxically, perhaps) make the problem easier to solve than it would have been without the constraints.

Reasoning in innovation

Most traditional idea-generation techniques (such as brainstorming) assume that people can use some form of reasoning to generate a creative solution to a problem. Brainstorming techniques are typically inefficient, and they often lead to fewer ideas than would be generated had the group members worked alone (e.g., Mullen, Johnson, and Salas, 1991). One reason for this inefficiency is that these techniques are not rooted in any theory of the way the mind works. Osborn's original brainstorming technique was focused more on group dynamics than on cognitive processing. While it is certainly important that group members feel free to contribute ideas without fear of criticism, these rules alone are not sufficient to lead people to generate creative ideas.

There is an assumption at the core of brainstorming, however, that clearly has merit. There are ideas in the heads of people engaged in innovation (or perhaps more broadly in the heads and in the environments of the innovators) that can be accessed and combined in ways that will lead to novel solutions to problems. That is, reasoning processes in innovation are aimed at finding ways to reuse existing ideas.

There are several core reasoning processes involved in innovation that are central to the research described in this book. In this section, we discuss the role of analogical reasoning, conceptual combination, and the influence of principles on innovation.

Analogical reasoning is the ability to see one domain as similar to another based on commonalities in the relations that hold between domains (Gentner, 1983; Hesse, 1966; Holyoak and Thagard, 1989). The domains need not have similar objects in them to be seen as analogically similar. For example, Chapter 5 by Linsey et al. discusses the role of analogy in engineering design. This chapter opens by pointing out that a fuel cell must solve the problem of distributing fluid throughout the fuel cell. One successful design for a fuel cell draws an analogy between the vein system in a leaf and a fuel cell. The vein system in the leaf provides an elegant solution for maximizing the contact of the fluid in the leaf with the surface area of the leaf's interior. The solution implemented in the fuel cell has the same basic structure as that of a leaf, though it clearly differs on the surface. The fuel cell is not a living organism composed of living tissue. The fuel cell is manufactured, not grown. Thus, the domains are analogous. Of importance, recognizing that a similar problem has been solved in a different domain allows the structure of the solution from one domain to be applied to the other.

For those not familiar with the technical domain of fuel cells, a second example of analogy will assist in elucidating this concept. Consider the problem of "designing a surveillance system to be dropped by lightweight unmanned aerial vehicles." These "sentinel" systems have military and civilian applications where video feeds need to be transmitted, but where the placement of the sentinel is difficult due to inaccessibility of the surveillance site. When dropping these systems, a key sub-problem is for the sentinels to self-orient, or right themselves. An analogous solution comes from self-righting children's toys such as Weebles (by Hasbro™, "Weebles wobble but they don't fall down") and Punching Clowns (or Bobo dolls). These toys do not share similarities with the sentinel device in terms of intent or overall purpose. After all, these toys are meant to entertain children. However, critical similarity relationships do exist: both the sentinel system and Weebles or Punching Clowns must obtain a certain orientation after being dropped or knocked down. It is the recognition of one or more key relationships that provides the link to using analogies to solve difficult problems. Figure 1–2 shows a sentinel system with a camera and antenna mass, where it is clear that the device has a wide base and low center of gravity, similar to the Weebles and Punching Clowns.

As many of the chapters in this volume make clear, the reuse of knowledge, as with the fuel cell and sentinel surveillance system, is a two-edged sword. On one hand, existing knowledge may be the source of profound

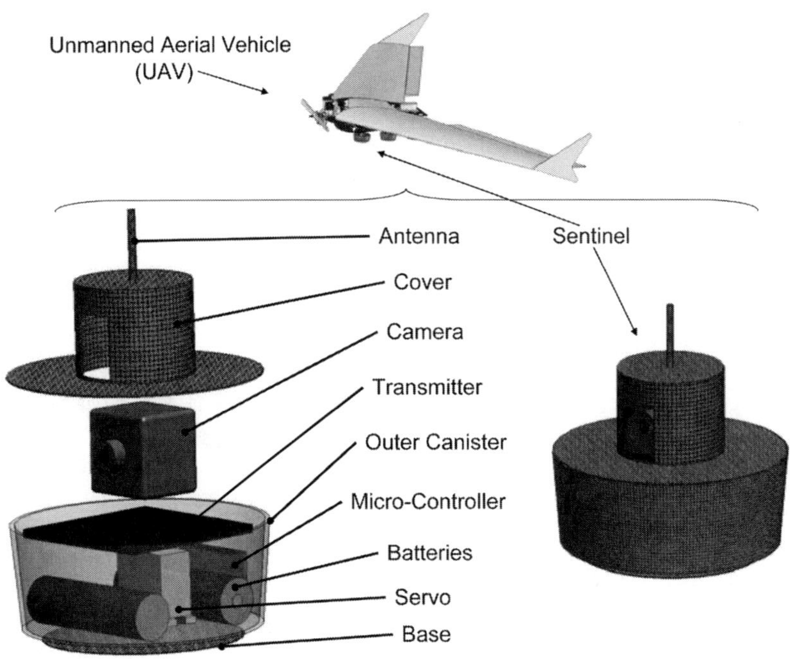

Figure 1–2 An example of a self-righting sentinel. The design is based on self-righting children's toys.

insight into difficult problems. On the other hand, designers often get fixated on a single idea and find it difficult to break free of that idea. The chapters by Christensen and Schunn (Chapter 3); Moreau and Dahl (Chapter 6); Smith, Kerne, Koh, and Shah (Chapter 7); and Ward (Chapter 8) all focus on this issue in different ways. For example, Ward points out that innovation often takes the path of least resistance. It begins by focusing on domain knowledge that is similar to the domain of the problem being solved. The retrieved knowledge is then tweaked or transformed to help solve the current design problem. However, by using existing knowledge from similar domains, new designs are often quite similar to existing designs. Similarly, Christensen and Schunn (2007) studied engineering design teams in the domain of medical plastics. They found that when the team was working with a physical model of a prototype product, they found it difficult to retrieve and use domain knowledge that went beyond that model. Instead, the domain knowledge used in this case was typically concretely similar to the prototype. Thus, an important function of tools for innovation is to help innovators to use prior knowledge without getting too strongly focused on a single instance.

One way that prior knowledge can be organized in order to avoid a focus on specific instances is to extract principles of design. A *principle* is a strategy for design that specifies relationships among items without focusing strongly on the objects themselves. The principle can then be applied to many different domains. In this way, a principle is like the concept of a *schema,* which is often discussed in the literature on analogical reasoning (Gick and Holyoak, 1983; Ohlsson, 1993; Schank and Abelson, 1977). One way that a principle differs from a schema is that principles are associated with specific instances that embody that schema. Thus, a principle serves both as an abstraction—i.e., a meta-analogy—but also as an organizing principle that enables designers to access relevant prior knowledge.

Chapter 9, by Singh, Walther, Wood, and Jensen, examines the role of principles in designs that involve transformations. For example, a sofa bed contains the elements of a bed that fold up in a way that stores them away most of the time, allowing the bed to function as a sofa. The sofa bed makes use of the transformation principle, or meta-analogy of "expand/collapse," which is a general component of many designs that involve transformations from one state to another. Singh et al. also present a tool for supporting designers who want to use transformation as part of a product design.

A third aspect of reasoning that is crucial for innovation involves people's ability to combine concepts. Analogical reasoning and principles alone cannot complete the innovation process. Recognizing that two domains are similar or that a particular principle may apply to a design problem does not end the innovation process. At that point, designers must combine the domains and principles to reach a solution. This conceptual combination process has been the focus of psychological research.

One difficulty in this kind of conceptual combination is determining what aspects of a prior problem should be carried over to a solution. In Chapter 7, Smith, Kerne, Koh, and Shah discuss ways that people may fixate on particular aspects of a known design in ways that may limit the effectiveness of a final solution. For example, they describe a task in which designers attempted to create a spill-proof coffee cup. They give an example solution that had a straw and a mouthpiece, but state that the final solution could not use these elements. They found that designers had difficulty generating designs that did not have these elements. In order to facilitate the combination of concepts, they present a computer program that enables designers to create a collage of relevant information sources. When designers compare across information sources, they are less likely to be affected by elements of particular examples. They demonstrate that their computer-based tool leads to more novel properties in an innovation task. This finding suggests that the use of many base domains to solve a target problem may be an effective way to help designers find relevant aspects of a solution from prior knowledge (Gick and Holyoak, 1983).

In addition, some representation formats may have inherent ambiguities that may further influence innovation processes. In Chapter 4, Tversky and Suwa examine the role of sketches in innovation. They find many instances in which elements of sketches may be reinterpreted to provide further insights into a problem. In this case, the innovation process relies on the inherent ambiguity of sketches to suggest additional novel aspects of a design.

Finally, causal reasoning is a crucial aspect of innovation. Developing an innovative solution to a problem requires knowing quite a bit about the causal relationships in that domain. Of course, design teams usually have a number of domain experts in them. However, there are two areas in which causal knowledge may be lacking. First, as Bridewell, Borrett, and Langley discuss in Chapter 11, innovation in science is directed specifically at pushing the boundaries of causal explanations. Their chapter describes a computational system that models the development of causal models in science.

Second, when design teams must use knowledge outside of the expertise of the members of the team, that knowledge may be sparse. Rosenblit and Keil (2002) find that the quality of people's causal explanations is often much lower than they expect it will be. Thus, even when a design team finds a promising analogy between the current problem and some other domain, it may be difficult to transfer that solution to the new domain without more details about the causal knowledge in the domain of the potential analogy. These gaps in people's knowledge suggest that another avenue for the development of tools for innovation is to find sources of causal knowledge about domains that may provide a potential solution to a new problem.

To summarize, there are several reasoning processes that are central to the innovation process. Chief among them is analogical reasoning, which helps designers reuse existing knowledge in innovation by allowing knowledge from one domain to be used to solve problems in another domain with a similar structure. Analogies have the potential limitation that people may become fixated on particular details from the analogous domain, so the extraction and use of principles is also an important aspect of innovation. Finally, causal reasoning is a crucial part of the innovation process. Many of the chapters in this volume focus on tools that relate directly to these reasoning processes.

A Cognitive Science Approach to Innovation

The chapters in this book reflect an interdisciplinary approach to the study of innovation. In our view, this interdisciplinary approach is crucial for three reasons. First, there are independent literatures on innovation across the cognitive sciences, and only interdisciplinary research will bring them together. Second, a deeper understanding of basic cognitive processes can illuminate design tools and design methods in fields outside of cognitive science. Third, domain knowledge is crucial to innovation, but most basic research in cognitive science tends to minimize the role of background knowledge. Thus, research on innovation that focuses on disciplines outside of cognitive science will also illuminate basic cognitive processes that have resisted study so far. We discuss these issues in more detail in this section.

Studying Innovation

It is no surprise that there is a literature on creativity within cognitive science. What may be surprising, however, is that the literature that is explicitly about creativity is not central to research in cognitive science. There is some research on the attributes of creative individuals (Simonton, 2000) and on the makeup of effective groups (Paulus and Nijstad, 2003). Research has explored the psychological states that lead to creativity (Csikszentmihalyi, 1996). There is some work testing the effectiveness of traditional methods of group creativity like brainstorming (Mullen, Johnson, and Salas, 1991; Paulus and Brown, 2002). In addition, there is some research on factors that influence the creativity of ideas that people generate (Finke, Ward, and Smith, 1992). There are specialized journals for research on creativity. However, little research that is explicitly about creativity is represented in the main journals in the field.

These references do not form an exhaustive list of studies of creativity, of course, but they bracket the range of topics that have been part of the psychology literature. Interestingly, much of this work is done on the fringes of the field rather than in the mainstream. Indeed, Ward, Smith, and Vaid (1997) present the results of a conference of mainstream psychologists whose research bears on creativity, but whose work is not directly focused on creative processes. This volume is interesting because of the number of psychologists who could potentially address issues of creativity quite easily, but who do not typically think of themselves as studying creativity. Thus, researchers from other disciplines who are interested in creativity will have a hard time finding relevant literature.

Other disciplines that rely on creativity all seem to have their own literatures on this topic. In disciplines that traditionally interact with cognitive science, there is a reasonable exchange of ideas. For example, there is work in the business literature that examines creativity in the workplace. This literature often draws from psychological research (e.g., Amabile, 1988). In many other disciplines, however, there is a long history of research on creative processing that makes little explicit contact with work in cognitive science.

The design literature in engineering has this quality. There are many heavily cited historical works in engineering design, such as *Introduction to Design* (Asimow, 1962), *Conceptual Design for Engineers* (French, 1971) and *Conceptual Blockbusting: A Guide to Better Ideas* (Adams, 1979) that focus on ideation. There is likewise a thriving research community within engineering whose aim is to develop methods to improve the process of generating new and innovative designs (e.g., Otto and Wood, 2001). Many techniques arise as an evolution of methods that have worked in practice, rather than a systematic merger of research in cognitive science and research in engineering design. For example, Shah et al.'s (2001) collaborative sketching (C-SKETCH) method elaborates on sketching methods that have traditionally been part of engineering methods.

One prominent design method, TRIZ, developed from an analysis of patents (Altshuller, 2005). This method identified 40 dimensions of variation that characterize successful innovations. For example, a successful design may take a product with a relatively undifferentiated surface and add texture to the surface to support additional functions. Designers are encouraged to analyze existing products to determine aspects of a product that could potentially be transformed using these design dimensions. This method is quite successful at helping designers to improve on existing products.

There are of course many other domains for which creativity is also central. Obviously, the fine arts prize creativity. Interestingly, there is little cross-talk between those who study artistic creativity and those who study basic cognitive processes. There has been some interest in the potential relationship between mental disorders and artistic creativity, but very little work on basic cognitive processes in normal individuals that may contribute to artistic creativity. This tension between art and science is also evident in architecture, which also prizes the creative artistic talents of its practitioners. Consequently, there is quite a bit of room for new research bringing together design research in applied disciplines with basic studies of creativity within cognitive science.

Bringing the Lessons of Cognitive Science to Other Disciplines

As it is a discipline of basic science, there is a general expectation that research in cognitive science will be used to advance our understanding of innovation and design in more applied domains like engineering and business. Consistent with this expectation, there is growing collaboration between innovation researchers in applied disciplines and those in the basic cognitive sciences.

Because this direction of information flow is the least surprising, we will say little about it here. However, many of the chapters in this volume do represent collaborations of this type. For example, Chapter 7 by Smith, Kerne, Koh, and Shah uses findings from laboratory research on memory to make predictions about the circumstances in which people will get fixated on a particular idea when creating a novel design. These predictions are then tested in applied research with engineers. Chapter 11 by Bridewell, Borrett, and Langley applies basic research in machine reasoning to an applied task of hypothesis-generation in science. Chapter 5 by Markman, Wood, and

Linsey applies basic research on analogical reasoning to the study of design. Finally, Chapter 8 by Ward applies basic research on categorization to further our understanding of ways people extend existing ideas to new situations. As these chapters represent, research on innovation recognizes the importance of basic cognitive processes and seeks to apply insights from research across the cognitive sciences to the study of design.

Advancing the Study of Cognitive Science Using Content

It is typical to think about knowledge from basic sciences being extended to applied disciplines. However, cognitive science has some special properties that allow much to be learned about basic cognitive processes from the study of applied disciplines. In particular, cognitive science aims to understand general cognitive processes that apply across domains. Consequently, cognitive science tends to focus on aspects of processing that are not tied to the content of particular domains. Instead, cognitive science tends to focus on aspects of the structure of knowledge or the structure of situations. We illustrate this point with a few examples.

Research on memory is usually focused on general aspects of the structure of knowledge and of memory situations. For example, studies often ask people to remember lists of words rather than aspects of their lives. Experiments then manipulate aspects of the way the information is presented. Research has examined influences of the task being done on later memory of items from a list (Tulving, 1983). For example, judging whether a word appears in uppercase or lowercase letters leads to worse memory than judging whether a word makes sense in a particular sentence. Research on false memories has used a paradigm in which the list contains the most frequent associates of a target word (but not the target word itself), and demonstrates that people often misremember the target word as being on the list (Roediger and McDermott, 1995). This research strategy makes a lot of sense, because memory research aims to make predictions about memory in general and not just memory for items from a particular domain.

Similarly, studies of categorization have focused on issues like whether people learn the specific category members they encounter, or whether they extract an average or prototypical member of a category (Medin and Schaffer, 1978; Posner and Keele, 1970). The particular content of the categories presented is less important to this research than the structure of the category members. Again, the idea is to make general claims about categorization rather than learning something about specific categories that people know (Markman and Ross, 2007).

As a third example, analogical reasoning focuses on general statements about mental representations such as making a distinction between the representational elements that describe objects and the relations that relate two or more aspects of a domain (Gentner, 1983, 1989). This work demonstrates that good analogies preserve similarities in the relationships between domains regardless of whether the objects in those domains are at all similar.

In each of these domains of cognitive science, much has been learned about basic cognitive processes from this research. In each of these areas, however, the work tends to focus on simple tasks that can be done in a relatively short experimental session (usually about an hour) using easily available people (like college students) as research participants. Consequently, little is known about how expertise influences basic processes like memory, categorization, and analogy. One reason that expertise is poorly understood is that experts know a lot about a specific domain.

This fact about experts presents two barriers to research. First, in order to do research with a particular expert population, it is necessary to work with experts in that domain. Thus, doing research with experts requires a significant investment of time, effort, and resources. Second, there is some risk involved with this research. After all, if our understanding of cognitive processes is sufficiently good, then research with experts will primarily serve to confirm what we already know about cognitive processing. Because of these two barriers to research, little research in cognitive science focuses on expertise in particular content domains.

Studies of innovation, however, require expert populations. Creativity and innovation involve domain experts. Indeed, one reason why basic cognitive science research on creativity and insight has not advanced as rapidly as studies of other cognitive processes is that it is difficult to study creativity in college students who have little or no specific expertise as a group.

Collaborations between basic researchers and researchers in applied disciplines can bear great fruit for basic research on creativity and for the understanding of cognitive processing more broadly. The beauty of examining innovation processes in domains like engineering is that the worst-case scenario (from a research standpoint) is that the research results in new design techniques and tools to support innovation in that area of expertise. More importantly, however, we may also develop important new insights into basic cognitive processes by examining the performance of experts.

Creating Tools for Innovation

Of course, the main function of this book is to examine the transition from studies of expert performance in innovation settings to the development of tools that support innovation. A tool for innovation is a cognitive prosthetic that somehow increases the capability of individuals and groups. It enhances the ability of problem solvers to generate and develop ideas beyond their innate or *ad hoc* processes. There are three broad classes of innovation tools that are represented in the chapters in this book. First, there are tools that extend the knowledge of individuals and groups to provide additional domains that may be useful for solving difficult design problems. Second, there are tools for affecting the content of what people are thinking about, to play on people's strengths at finding connections between domains. Third, there are tools for structuring the design process to make the work of individuals and groups more systematic.

Tools for extending the knowledge of individuals and groups can come in many forms. One is to create systematic databases of known solutions to frequent design problems. These databases or repositories are particularly important in large organizations that have proprietary knowledge that may be useful across different units within the organization. For example, Chapter 10 by Summers, Anandan, and Teegavarapu describes tools of this type. A key problem in creating these databases is developing procedures to allow the database to be searched to find the relevant knowledge. This problem is particularly important in situations where the problem that was initially solved shared only relational similarities to the current problem. Chapter 5 by Markman, Wood, and Linsey discusses tools that seek analogical matches to a database query.

At present, of course, humans are more skilled at making connections among domains than even the best machines. Thus, another class of tools tries to maximize the strengths of people's ability to make connections by influencing the information that people think about. For example, Chapter 3 by Christensen and Schunn discusses a tool that provides random information about near and distant domains to cue memory for information relating to the cue. Being presented these cues at strategic points in the innovation process, the designer may be spurred to make connections that might not occur during the normal course of processing.

Similarly, Chapter 8 by Ward discusses a computer-based tool that organizes related concepts to spur people to reconceptualize a problem. In particular, these tools would provide both a more general description of the problem being solved as well as other more specific domains that also solve the same problem. These more specific domains provide other avenues for making connections between the current problem and other knowledge.

As a third example, Chapter 7 by Smith et al. discusses a program called combinFormation that allows users to organize knowledge drawn from databases and the Internet into a collage. The purpose of this tool is to promote juxtapositions of concepts that are not normally conceptualized together as a way of breaking out of impasses.

Finally, tools can help make the design process more systematic. For example, as discussed above, Chapter 9 by Wood et al. analyzes products that transform and identifies three basic principles that guide the creation of transformers, as well as a series of specific methods for implementing those principles. From this analysis, they have created a deck of cards that can be used during the design process to employ these principles systematically to create new products that can change their state.

Moving Forward

As the chapters in this book demonstrate, there is renewed interest in using the cognitive science of innovation to develop tools to increase the effectiveness of innovation processes. We see two key avenues for growth of this research in the future. First, it will be important to promote more discussion among cognitive scientists and design practitioners across the range of design activities, including areas like architecture and fine arts that often have less contact with cognitive science than fields like engineering. Collaborations like those represented in this book are at the front end of a growing interest in innovation and design methods.

As this work matures, however, it will also be important to involve actual innovation teams in the process of tool development. Just as the collaboration between cognitive scientists and researchers in applied disciplines can suggest new questions for basic research, application of tools in real design settings suggests new aspects of the innovation process that may be missed by focusing primarily on more-academic research. For example, in large organizations, there is a long distance between a group meeting to generate design ideas and the actual implementation of that design. Often, good ideas get lost along this journey. By tracking innovations through real organizations, future research can focus on tools to go all the way from idea generation to completed designs. Ultimately, the success of tools for innovation must be judged by the degree to which they facilitate real innovations in real settings.

References

Adams, J. (1979). *Conceptual blockbusting: A guide to better ideas.* New York: W.W. Norton.

Altshuller, G. S. (2005). *40 principles: TRIZ keys to technical innovation.* Worcester, Mass.: Technical Innovation Center, Inc.

Amabile, T. M. (1988). A model of creativity and innovation in organizations. *Research in Organizational Behavior, 10,* 123–167.

Asimow, M. (1962). *Introduction to design.* Englewood Cliffs, N.J.: Prentice-Hall.

Boden, M. A. (1994). What is creativity? In M. A. Boden (Ed.), *Dimensions of creativity.* Cambridge, Mass.: The MIT Press.

Christensen, B. T., and Schunn, C. D. (2007). The relationship of analogical distance to analogical function and pre-inventive structure: The case of engineering design. *Memory and Cognition, 35*(1), 29–38.

Csikszentmihalyi, M. (1996). *Creativity: Flow and the psychology of discovery and invention.* New York: HarperCollins.

Finke, R. A., Ward, T. B., and Smith, S. M. (1992). *Creative cognition: Theory, research, and applications.* Cambridge, Mass.: The MIT Press.

French, M. (1971). *Conceptual design for engineers.* Springer-Verlag, Berlin.

Gentner, D. (1983). Structure-mapping: A theoretical framework for analogy. *Cognitive Science, 7,* 155–170.

Gentner, D. (1989). The mechanisms of analogical learning. In S. Vosniadou and A. Ortony (Eds.), *Similarity and analogical reasoning* (pp. 199–241). New York: Cambridge University Press.

Gick, M. L., and Holyoak, K. J. (1983). Schema induction and analogical transfer. *Cognitive Psychology, 15*(1), 1–38.

Hesse, M. B. (1966). *Models and analogies in science.* South Bend, Ind.: University of Notre Dame Press.

Holyoak, K. J., and Thagard, P. (1989). Analogical mapping by constraint satisfaction. *Cognitive Science, 13*(3), 295–355.

Markman, A. B., and Ross, B. H. (2007). *Categories in use.* San Diego, Calif.: Academic Press.

Medin, D. L., and Schaffer, M. M. (1978). Context theory of classification. *Psychological Review, 85*(3), 207–238.

Mullen, B., Johnson, C., and Salas, E. (1991). Productivity loss in brainstorming groups: A meta-analytic integration. *Basic and Applied Social Psychology, 12*(1), 3–23.

Ohlsson, S. (1993). Abstract schemas. *Educational Psychologist, 28*(1), 51–66.

Otto, K., and Wood, K. L. (2001). *Product design: Techniques in reverse engineering and new product development.* Upper Saddle River, N.J.: Prentice Hall.

Paulus, P. B., and Brown, V. R. (2002). Making group brainstorming more effective: Recommendations from an associative memory perspective. *Current Directions in Psychological Science, 11,* 208–212.

Paulus, P. B., and Nijstad, B. A. (Eds.) (2003). *Group creativity: Innovation through collaboration.* New York: Oxford University Press.

Posner, M. I., and Keele, S. W. (1970). Retention of abstract ideas. *Journal of Experimental Psychology, 83,* 304–308.

Roediger, H. L., and McDermott, K. B. (1995). Creating false memories: Remembering words not presented in lists. *Journal of Experimental Psychology: Learning, Memory, and Cognition, 21* (4), 803–814.

Rosenblit, L., and Keil, F. C. (2002). The misunderstood limits of folk science: An illusion of explanatory depth. *Cognitive Science, 26,* 521–562.

Schank, R. C., and Abelson, R. (1977). *Scripts, Plans, Goals and Understanding.* Hillsdale, N.J.: Lawrence Erlbaum Associates.

Shah, J. J., Vargas-Hernandez, N., Summers, J. D., and Kulkarni, S. (2001). Collaborative sketching (C-Sketch): An idea generation technique for engineering design. *The Journal of Creative Behavior, 35*(3), 168–198.

Shaw, G., Brown, R., and Bromiley, P. (2002). *Strategic stories: How 3M is rewriting business planning.* Cambridge, Mass.: Harvard Business School Press.

Simonton, D. K. (2000). Creativity: Cognitive personal, developmental, and social aspects. *American Psychologist, 55,* 151–158.

Tulving, E. (1983). *Elements of episodic memory.* New York: Oxford University Press.

Ward, T. B., Smith, S. M., and Vaid, J. (Eds.) (1997). *Creative thought: An investigation of conceptual structures and processes.* Washington, D.C.: American Psychological Association.

CHAPTER 2

ON "OUT-OF-THE-BOX" THINKING IN CREATIVITY

ROBERT W. WEISBERG

Creativity has become the critical element in the survival of the modern corporation (Kelley, 2000). Corporations must adapt to a constantly changing environment through the development of novel products—through innovation—or they will not survive. Therefore, creative thinking—the thought process that brings about novel ideas and, *ipso facto*, the thought process that underlies innovation—is crucial for the survival of the modern corporation. It then becomes important to understand creativity; i.e., to determine the process whereby novel ideas are brought about. If we could understand that process, we could create methods to help R&D departments develop the new products on which their companies and, ultimately, the world's economy depend. This chapter presents an analysis of the thought process underlying creativity, as developed through empirical studies of the creative process, in order to provide a foundation of data to support discussions of methods that foster innovation.

Before discussing some of the details concerning the creative process that have arisen from the research (e.g., Weisberg, 2006), let's define the concepts we will be dealing with, among which are *creativity*, *innovation*, *design*, and *invention*. There are many closely related concepts involved, so it is important to tease them apart. This chapter explores a cognitive psychologist's perspective on creativity, and so it will only fit in varying degrees with what researchers in other fields think about the concepts involved (indeed, what is written here may not fit with what other cognitive psychologists think about those concepts). However, a set of working definitions will at least provide us with a platform from which we can begin to explore commonalities and differences in thinking.

Definitions of Relevant Concepts

Creativity

Creativity entails the production of goal-directed novelty (Weisberg, 2006, Chap. 2). Creativity results in the intentional production of new things, either ideas or physical objects; the *creative process* or *creative thinking* is the psychological means whereby such novelty is brought about. Assuming that the individual's intention is critical in creative production means that one cannot be called "creative" if one produces something new by accident. The subsequent utilization of that accidental novelty might involve processes that we could label as creative. The initial "discovery" did not, according to the definition assumed here, come about through the creative process. It is generally not useful to include *value* in this definition. (This assumption goes against most definitions of creativity; see, e.g., chapters in Runco and Pritzker, 1999; and Sternberg, 1999, for many examples.) Defining creativity as the production of novel products that are of value (no matter how one defines *value*) results in complexities that render the definition unusable (for discussion, see Weisberg, 2006, Chap. 2). Most important, the value of some product can change over time, which means that, if we include value in our definition of *creative*, the products or persons that one generation classifies as creative might not be so classified by the next. That possibility means that our database would be constantly shifting as we tried to develop our understanding of creativity and related concepts—an unacceptable set of circumstances.

In sum, this chapter is concerned with the issue of how new things are intentionally brought about, whether or not those things turn out to be useful in any way. With this perspective, we would be just as interested in the psychological processes underlying the development of a new airplane that never got off the ground as in those underlying one that did.

Innovation and Design

Innovation involves a product that meets some criteria beyond those of intention and novelty; an *innovation* is a new product that serves some purpose. It is here that questions of value become important. Building on the definition of creativity just given, an innovation is a new product that was intentionally produced to serve some purpose and that succeeds, to a degree that is acceptable, in doing so. *Design* is the process whereby innovation is brought about. So the design process encompasses creativity (the generation of novelty) plus something more (the adjustment of that novelty so that it serves some specific purpose). See Figure 2–1 for the relationships among those concepts.

Invention

Innovation, design, and invention are closely related concepts; an *invention* is also a novel product that has been intentionally developed to serve some purpose (that is, an invention is also an innovation as defined above). But an invention is the *first* innovation within some class of objects. In other words,

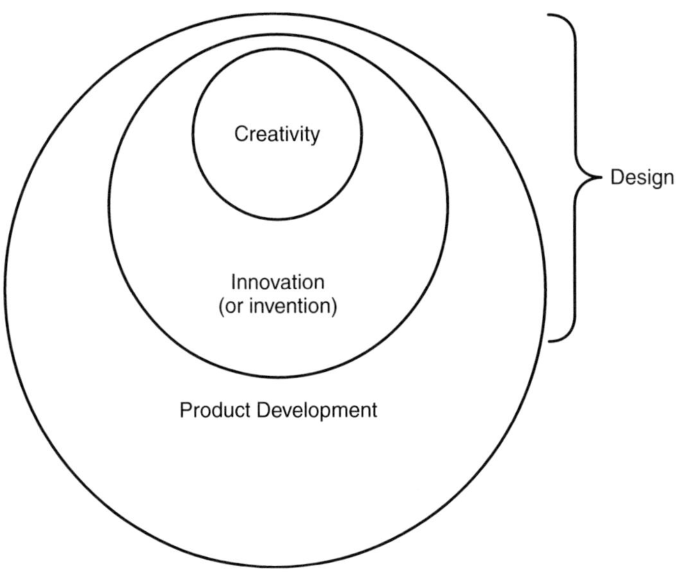

Figure 2–1 Relations among concepts used in this chapter.

a new member of an already existing category of objects is an innovation, but the first of the objects within that category is an invention. So, for example, it seems reasonable to say the following of the Wright brothers: (1) they invented the airplane; (2) they designed the first airplane; and (3) they designed an airplane. The individuals who successfully followed the Wrights only succeeded in designing airplanes. Those individuals may have invented things in their work—components of their successful airplanes—but they did not invent the airplane. The airplane—or any artifact—can only be invented once.

Thus, the processes of invention and innovation might be the same, except for the fact that the former results in production of the first of some class of objects (i.e., the first airplane) while the latter results in additional members of the already populated category (i.e., other airplanes). It is an empirical question as to whether the process of invention is the same as the process of innovation. That is, is the same process (or processes) involved in producing the first member and subsequent members of some class of objects?

Marketing and Product Development

There is another step worth making explicit here: the successful marketing of an innovation. It is one thing to produce an innovation (or an invention); it is quite another to successfully market that product. *Product development* is the successful marketing of an innovation.

Assumptions about Creativity: Out-of-the-Box Thinking

One could say that creative ideas come about through the rejection of old ideas, as the creative thinker makes a leap to the new. This perspective is so familiar to us all that it has become part of our common culture. When one talks about "thinking outside of the box," the "box" that our thinking must break out of is formed by the constraints brought about by the old ways of looking at things. We box ourselves in through the limitations we put on ourselves by our past experience, which constrain the ways we can think. Thinking inside the box puts us at a disadvantage when we are in situations that demand novelty. This view is also explicitly accepted by psychologists who study creativity (e.g., Frensch and Sternberg, 1989; Simonton, 1995). Given this perspective, it follows that if we wish to be creative—to be innovative—we need to develop ways to break out of the box. We need to break out of the constraints we unwittingly place on ourselves when we rely on our old ways of doing things in situations that demand that we come up with something new.

Although it is critical that we develop methods to facilitate creative thinking, such a discussion may be considered by some to be beside the point, because there are already extant methods that have been designed especially to allow us to go beyond the bounds of ordinary thinking. The most well-known of those methods is brainstorming (Osborn, 1953), which was developed to facilitate the production of the largest number of new ideas possible in a situation. The attempt is made to have people throw off the constraints that typically bind idea production, and to bring to the fore ideas that would typically never see the light of day under ordinary thinking. Furthermore, there are companies—*idea factories*—that are willing to take on the task of producing novel ideas for clients (Kelley, 2000). Presumably the idea factory knows things about methods for producing novel ideas, such as brainstorming, that the client does not know. Although the staff of the idea factory may not work in the specific industry of the client corporation, it is assumed that they know enough about creative thinking to produce novel products within that industry.

General Methods of Thinking Creatively

The assumption that an idea factory can serve to produce creative works in a domain in which its employees might not possess expertise is based on a prior assumption about creative thinking: The methods used to produce novel ideas are not dependent on the domain in which one is working; that is, they are domain-general methods. Indeed, if it is assumed that creative production depends on thinking outside of the box—on breaking away from the constraints put on us by past experience—then it is true by definition that the production of novel ideas is independent of knowledge of the discipline. One may be better off if one knows only a minimal amount about the domain in which there is a problem to solve. One should know enough to navigate within the domain, but not enough to become trapped in the limitations that experts, perhaps unknowingly, bring with them (Frensch and Sternberg, 1989; Simonton, 2003).

These questions raised—How can we understand creative thinking? How can we use our knowledge of the creative process to develop methods of fostering innovation?—already have been answered. But to a cognitive scientist studying creative thinking, there is still a basic problem with all the ideas presented so far: The notion that creativity depends on out-of-the-box thinking is a myth. The components of what can be called "the out-of-the-box view" (creativity depends on out-of-the-box thinking; production of new ideas depends on rejection of the past; expertise can put constraints on creativity; there are general methods for developing new ideas) may make sense, but is there evidence that they are true? A review of evidence from studies of creative thinking, presented in the next section, will provide no support for the idea that creativity depends on out-of-the-box thinking.

Empirical Evidence Concerning Workings of the Creative Process

One way to determine if creative thinking comes about through out-of-the-box thinking is to study historically important creative advances—case studies of seminal creative achievements—to determine if they indeed did come about as a result of rejecting the past. So let us consider several examples of creative thinking at the highest level to see if they have involved going outside of the box. The case studies to be examined include:

- Watson and Crick's discovery of the double helix
- The Wright brothers' invention of the airplane
- Edison's invention of the kinetoscope (the first moving pictures)
- Picasso's creation of his great painting *Guernica*
- A case study of innovation in industry: IDEO's development of a new shopping cart

We will see that those seminal advances, in science, in technology, in the arts, did not involve out-of-the-box thinking. Contrary to the perspective that is taken for granted by many, those creative advances were brought about through building on the past, rather than rejecting it. Furthermore, the analyses of case studies will also indicate that the thought processes underlying creative advances are the same "ordinary" thought processes that we use in our day-to-day intercourse with the world as professionals and as ordinary folks. Examples of those ordinary processes are retrieval and use of information from memory; logical thinking, both inductive and deductive;

analogical thinking; and analysis and comprehension of verbal and pictorial materials. Those conclusions, if valid, mean that we should rethink our expectations about how we should attempt to foster innovation.

The Double Helix

In the spring of 1953, James Watson and Francis Crick published their model of the structure of DNA, which went on to have a revolutionary effect on all our lives. Watson and Crick were only one of several groups of investigators who were trying to determine the three-dimensional structure of the molecule (for details on Watson and Crick, see Olby, 1972; and Judson, 1979; see also Weisberg, 2006, Chap. 1). It was assumed that knowing the structure of DNA would allow us to understand how the genetic material was passed on across generations and also to understand how the genetic material played the role of directing the metabolism of the cell. Those assumptions have been shown to be correct, as our knowledge of the structure of DNA has led to a new and more profound understanding of the processes underlying life.

Early in Watson and Crick's collaboration, they decided that they would *assume* that the DNA molecule was helical in structure, and from then on they looked at all the available evidence from the perspective of what it would tell them about the parameters of the helix. Several other investigators, who were, like Watson and Crick, highly qualified researchers in molecular biology, did not make that critical helical assumption, and so were left to attempt to work out the specifics of the structure through very slow and tedious methods. This "helical" decision was one factor in Watson and Crick's being the first to propose the double helix model of DNA. Thus, one might be tempted to say that Watson and Crick went outside of the box when they made the initial assumption that DNA was a helix, since no experimental evidence then available unambiguously pointed to a helical structure. Watson and Crick made that assumption based on something other than logic: one could call it intuition. So the critical question becomes: whence did that intuition arise?

Analogical Transfer

Watson and Crick's critical assumption or intuition came from the work of Linus Pauling, a world-famous chemist who had recently proposed a helical structure for the protein alpha-keratin, which forms hair, nails, and animal horn, among other components of organisms. Based on their knowledge of Pauling's work—in other words, based on their *expertise* as molecular biologists—Watson and Crick decided that it was reasonable to assume that DNA was also helical. Why did they make that critical decision? DNA and alpha-keratin are *analogous* molecules: their structures are similar in a number of ways. Both are: (1) organic molecules; (2) very large; (3) made up of basic building blocks—protein: amino acids, DNA: nucleotides; and (4) the building blocks are linked one to the next through the same type of bonds. So Watson and Crick's intuition did not come about through out-of-the-box thinking: they did not reject the past, they did not reject what they knew. They used the past as the basis for constructing their creative advance. Returning to the question of the role of ordinary thinking in creativity, Watson and Crick used information from a domain closely related to the one they were working on—they used a *near analogy*—in order to make progress. They did not make a leap outside the box to some remote domain in order to begin to understand DNA. Their "intuition" was a move *inside the box*.

It should be noted that there was still creativity involved in Watson and Crick's advance, even though that advance was solidly based on the past, since deciding that DNA was a helix was only the first step in their discovery. They had to go beyond Pauling, because the double helix of DNA is not identical to Pauling's alpha helix. The protein structure is a one-strand helix, in contrast to the two-strand helix of DNA. Thus, Watson and Crick had to determine the specifics of the DNA structure, which took a year and a half. However, Watson and Crick got a leg up on their problem by incorporating Pauling's work as the foundation of their own. Furthermore, Watson and Crick filled in the details of the structure of DNA on the basis of information—research results—obtained from analyses of DNA by themselves and others. That is, ordinary scientific thinking—interpretation of research results, including logical reasoning—played a critical role in allowing Watson and Crick to achieve the goal of formulating a plausible model of the structure of DNA.

Watson and Crick did not use outside-of-the-box thinking in order to decipher the structure of DNA. They were working within a box: within their structured expertise as molecular biologists and, more generally, their expertise as scientists. They were using the ordinary thought processes of expert scientists, and the world was comprehensible through that expertise. They assumed that DNA was like other organic molecules, and they were correct: the puzzle yielded to their expertise. In considering the role of expertise in the discovery of DNA, it is illuminating to ask the following question: Could a non-expert have done what Watson and Crick did? The answer seems obviously to be no.

The Wright Brothers and the Invention of the Airplane

In December 1903, after some four years of work, Wilbur and Orville Wright carried out the first powered flight (for details on the Wrights, see Heppenheimer, 2003; and Weisberg, 2006, Chap. 5). A number of other research teams in the United States and Europe were also working on the problem of powered flight, but the Wrights were the first to achieve that goal. One critical difference between the Wrights and the other investigators was that the Wrights incorporated a *control system* as part of their flying machine. Other researchers assumed that the pilot would not play an active role in the control of the airplane in flight, because they believed that the pilot would not be able to respond quickly enough to unpredictable wind gusts. Those researchers therefore attempted to structure their flying machines in various ways, for example by attaching the wings at an angle to the frame and by hinging the wings and tail, that would allow the airplane to respond automatically to changes in wind speed and direction and thereby take the human out of the equation.

The Wrights, in contrast, assumed that in order for successful flight to occur the pilot would have to be able to control the machine once it got into the air. They spent years on the beaches at Kitty Hawk on the Outer Banks of North Carolina, flying gliders of ever-increasing sizes to develop and perfect their control system. Only when the control system was perfected—that is, when they were confident that they could control their gliders in flight—did they think about adding an engine to their machine. The researchers who were working on automatic mechanical components found after years of frustration and failure that it was not possible to control a flying machine automatically. Thus, the Wrights succeeded where others, some of whom started before the Wrights and had much more in the way of resources, did not. So again we have an example of an intuition—the Wrights' need for a control system—and again we are faced with the question of where that intuition came from. As with Watson and Crick, the Wrights' intuition and its realization did not come about through thinking outside the box. The Wrights' knowledge—their expertise—served as the basis for their intuition and for its realization. We also have a second question of interest in the case of the Wrights: Once they had decided that they needed a control system, how did they develop the specifics of that system? It is one thing to decide that one needs to control a vehicle in flight and another to decide how that system will be realized. Here again we will see that the Wrights' knowledge about flying machines played a critical role in determining what the specifics of their control system would be.

On the Need for a Control System

In August 1895, Otto Lilienthal, a German engineer who had been experimenting with gliders as the basis for development of an airplane, was killed in a crash when his glider was caught in a gust of wind. Lilienthal's death was reported around the world, and the Wrights, who had not yet begun their work, read about it. Lilienthal's gliders were designed so that he hung from a bar underneath the glider's bat-shaped wing, somewhat like modern hang-gliders do. Lilienthal attempted to control his glider by swinging his body, thereby changing the center of gravity of the machine + pilot. However, that method was ineffectual in strong winds, as his death tragically demonstrated. The Wrights took Lilienthal's death, and others' who died while gliding, as evidence that control by the pilot was critical to successful flight. Other researchers did not trust the human pilot to be able to control an airplane in unpredictable gusts of wind, so they attempted to design mechanical systems that would remove the pilot from the equation. Why did the Wrights believe, contrary to other researchers working on the problem of powered flight, that a human could control an airplane? Where did that crucial intuition come from?

The Wrights' intuition that a human could control an airplane came from their expertise with bicycles (Heppenheimer, 2003). One often hears the Wrights described as "bicycle mechanics," but they were much more than that. They had designed and built bicycles, cutting and welding frames, for example, for a number of years in their own shop. The bicycle is critical in the story of the airplane because the bicycle and the airplane are similar as mechanical systems: When one makes a turn while riding a bicycle, one disturbs the dynamic equilibrium of the system. That is, when one turns on a bike, one points the wheels in the direction in which one wishes to go and then one *leans into* the turn; a turn on a bicycle is a controlled fall. One keeps from falling by maintaining one's speed. When one observes a learner on a bicycle making a turn, one sees that he or she typically does not go fast enough through the turn, and so must put down the inside foot to prevent the bike from falling over. Similarly, an airplane making a turn is in a roll, which is also a controlled fall. One tilts the wings and turns around the lower wing. If one does not maintain one's speed, the airplane will slide down along the path of the lower wing and spiral down from the sky.

One could say (and the Wrights seemed to believe) that an airplane is a bicycle with wings (Heppenheimer, 2003). A bicycle and an airplane are analogous and, as with DNA and alpha-keratin, they are near analogies: both are mechanical transportation devices that operate through dynamic equilibrium. The Wrights, similarly to Watson and Crick, did not go outside the box; there was no wide-ranging intuitive leap that went far beyond what they knew. They simply had knowledge that was applicable to the problem that they were facing.

Thus, the Wrights' experience with bicycles and the control of a dynamic system led them to the belief that they could learn to control an airplane. It should also be noted that the Wrights were not the only ones who saw the relationship between bicycles and airplanes. James Means, a commentator on the airplane scene, wrote, in an article that the Wrights probably read, that the airplane would be perfected by "bicycle men," because flying is like "wheeling" (Heppenheimer, 2003). Other researchers conceived of the airplane as a boat—part of an airplane is called a *rudder*—and that view turned out not to be useful. An airplane in a turn is not like a rowboat making a turn.

Development of the Control System

There are two questions related to the Wrights' control system: (1) Why did they feel a need for one? That need arose from Lilienthal's death combined with their knowledge of bicycles. (2) Once they decided that they needed a control system, how did they develop it? The Wrights' control system centered on moving the ends of the trailing edges of the wing tips in opposite directions (left-up, right-down; and vice versa). When the tips were tilted in opposite directions in that way, the glider or flyer would turn one way or the other. We thus have another "intuition" to analyze: where did the Wrights get the idea of bending the wing tips—which they called "wing warping"—to turn their aircraft? The wing-warping idea came from their reading concerning bird flight as well as their own observations of birds gliding essentially motionless on air currents (Heppenheimer, 2003). They noted that when a gust pushed the bird away from level, the bird would adjust its wing tips and then go back to a level position. They made a mechanical system to mimic the birds' movements and refined it, by making it lightweight and easy to adjust, so it would serve on their gliders. Then, as the result of years of practice, they learned to use it to control the orientation of their gliders in the air. They then transferred that system to their flying machine.

Wright Brothers' Airplane: Conclusions

Once again we have an example of what one could call a creative intuition, and once again we see that it did not come about through out-of-the-box-thinking. The Wrights took the idea of the need for a control system, and their belief that a human could control a flying machine in flight, from their experience with bicycles. The specific design of the control system came from another flying machine (albeit a living one), and they used it as the basis for their artificial system. Again expertise (transfer based on a near analogy) served as the basis for producing a creative response to a problem.

Edison's Kinetoscope

In 1879, Edison invented the phonograph, which made him world famous as the "Wizard of Menlo Park." (For more details, see Weisberg, 2006, Chap. 5.) The design of the phonograph was built out of several components (see Fig. 2–2A). First, a sound source was "captured" by a funnel. A membrane at the bottom of the funnel vibrated in response to the sound. That vibration was transferred to a needle attached to the membrane. The vibrating needle incised a groove on the surface of a spinning horizontal cylinder that moved below it. That incised groove corresponded to the sound source; that is, the groove was a record of the sound source. At playback, the process was reversed. The needle was placed in the groove

Thomas A. Edison's sketch of the phonograph. The final version of Edison's kinetoscope. Sketch of early motion picture device made by Edison. (Reproduced through the courtesy of the Edison National Historic Site.)

Figure 2–2 Edison's phonograph and kinetoscope. A. Early sketch for the phonograph. B. Final version of Edison's kinetoscope C. Early version of motion picture device.

on the cylinder; the cylinder was rotated, and the needle moved in response to the incised undulations in the groove. The membrane attached to the needle vibrated, sending out waves, and sound was heard at the wide part of the funnel.

In 1888 Edison invented the kinetoscope, the first device for presenting moving pictures. Figure 2–2B shows the final version of the kinetoscope. The machine is enclosed in a cabinet with an eyepiece at the top. A long strip of film, containing a series of pictures taken of a moving object over a short period of time, was drawn under the eyepiece frame by frame. A shutter inside the eyepiece opened and closed as each frame passed. A person looking through the viewer saw a moving image.

The development of the kinetoscope leaves us with the question of where that device came from. The final version of kinetoscope looks nothing like the phonograph, so we can see no relationship between the two devices. If, however, we examine an *early* version of the kinetoscope, we can see a striking relationship between that device and the phonograph. In the early version of the kinetoscope (see Figure 2–2C), visual information (a series of pictures) was attached in a spiral around a horizontal cylinder. At playback, the cylinder rotated; an eyepiece moved over the series of pictures; a person looked through the eyepiece; and moving pictures were seen. Thus, consideration from a historical perspective indicates that the kinetoscope was built on the phonograph.

In the patent application that he filed for the kinetoscope, Edison made clear the relationship between that invention and the phonograph: "I propose to do for the eye what the phonograph does for the ear. . . ." As with the other advances discussed here, the kinetoscope and phonograph are analogous. They are both communication devices, in which information is extended over time, and input and output are mirror-images of each other. Furthermore, as Figures 2–2A and 2–2C make clear, it is possible to present these different types of information in ways that are very similar on a physical level. In a parallel to Watson and Crick and the Wright brothers, Edison did not go outside the box when he invented the kinetoscope; he used a near analogy as the basis for the new device.

Picasso's Creation of Guernica

We have discussed several case studies in science and invention, and have concluded that several seminal creative advances have come about through what one could call "in-the-box thinking." One does not see great leaps of thought, which leave the known in their wake. Rather, the new is firmly built upon the foundation of the old. But does the same sort of "in-the-box thinking" occur in the arts? The arts are obviously different in content from science and invention, and perhaps in the thought processes involved. Therefore out-of-the-box thinking might be relevant in the arts. In Picasso's creation of *Guernica*, his great antiwar painting of 1937 (for additional discussion, see Chipp, 1988; and Weisberg, 2004, 2006, Chap. 1), out-of-the-box thinking did not play a role. (For discussions of other case studies in the arts, science, and invention, see Weisberg, 2006, Chap. 5.)

Picasso painted *Guernica* in response to the bombing of the town of that name by the Luftwaffe, the Nazi air force, which was in alliance with Franco in the Spanish Civil War. The town was not seen as being of particular strategic importance, and the bombing, which resulted in the destruction of much of the center of the town and 250 to 300 deaths, was seen by many as a terrorist tactic. The well-known painting contains the following characters when viewed from left to right: A bull stands with its body facing in toward the center of the painting, but its head is turned away. Below the bull, a woman, her head thrown back in a scream of agony, holds a dead baby, its head lolling backward. A broken statue, holding a broken sword and a flower in its hand, lies at the feet of the mother and child. In the center, a horse, its head up in a scream of agony, is dying from a lance that has pierced its body. At the top-center of the painting, above the head of the horse, are an electric light shaped like a human eye and a woman—on fire—leaning out of the window of a burning building, holding a light that also illuminates the scene. Below that woman, another woman with bared breasts is running into the scene from right to left. Finally, on the far right, another woman, also on fire, falls from a burning building with her arms stretched above her head.

For the student of creative thinking, *Guernica* is a very useful case, because Picasso numbered and dated the preliminary sketches he produced while working on it. Most important for the present discussion, the very first sketch he produced, dated 1 May 1937, four days after the bombing, is a sketch of the overall composition in which one can see the essence of the completed work. The light-bearing woman is in the upper center, leaning out of a building; the bull is on the left; and there is an object in the lower center that seems to represent the horse. Other composition sketches done on that day also contain the basics of the final composition. Thus, if we see the essence of *Guernica* in Picasso's first day of work, we can say that from the very beginning Picasso had an intuition of what he was going to produce. So here again we are faced with the question of where Picasso's intuition came from: Did Picasso think "outside of the box" in creating *Guernica*? In a parallel to the cases already analyzed, the answer to that question is no: *Guernica* was firmly based on Picasso's past, and he built on the past rather than rejecting it in creating his new work.

An example of a work from Picasso's earlier career that may have served as the basis for *Guernica* is the etching *Minotauromachy*, created in 1935, some two years before *Guernica* (Chipp, 1982; Weisberg, 2006, Chap.1). Myriad correspondences exist between the two works (see Table 2–1): both contain a bull (or bull-like organism), a woman holding a light, a dead person, a sword, a horse, birds, and a "vertical" person along one edge. It seems that either *Guernica* was built on the platform of *Minotauromachy*, or that both works were developed out of a common theme. Thus, the question of whether Picasso thought outside the box when creating *Guernica* can be answered in the negative. For further discussion, see Weisberg and Hass (2007).

Table 2–1 Corresponding elements in Picasso's *Guernica* and *Minotauromachy*.

Minotauromachy	Guernica
Bull (Minotaur)	Bull
Horse—head raised	Horse—head raised (stabbed—dying)
Dead person	Dead person (dead child—see below—and broken statue)
Sword (broken—in statue's hand)	Sword (in Minotaur's hand)
Flowers (in girl's hand)	Flower (in statue's hand)
Two women above observing woman on ground holding light	Woman above observing+holding light
Birds (standing in window above)	Bird (flying up toward light)
Vertical person (man fleeing)	Vertical person (burning woman falling)
Sailboat	Electric light
	Mother and dead child
	Woman running in

Conclusions—On the Origins of New Ideas

In all the cases discussed so far, in science, technology, and the arts, seminal creative advances came about through the building of the new upon the foundation of the old. There was not a wholesale rejection of the old; no "breaking out of the box" was seen. Each of those advances came about as a result of the creator's *staying within the box*: using expertise and domain-specific knowledge to go beyond what was known to produce something new. One important conclusion that arises from that analysis is that when it seems to us that a person has gone from the present to something completely new, that perception of an unfathomable leap of insight is the result of our ignorance. We do not know what the creative thinker knows, so, without a road map, we cannot follow the thought processes that led to the creative advance. When we get inside the creative thinker's head, however, we see that the new advances have been firmly built on the old, and that the thought processes are no different than the ordinary processes that we all use in our day-to-day interaction with the world.

On Innovation in the Corporation

The case studies presented so far draw from science, invention, and the arts. The findings have been consistent in demonstrating that novel ideas come about as the result of building on what one knows through the use of ordinary thinking. We have not found leaps outside of the box, away from what one knows. But are the findings relevant to the question of innovation in the corporate domain? Let us turn to an example of a well-publicized case study of corporate innovation, in order to demonstrate that the same processes are involved there. The example to be considered is IDEO's development of a new shopping cart, which is discussed in detail by Kelley (2000).

IDEO was given a challenge by Ted Koppel on his popular *Nightline* late-night program: to develop an improved shopping cart. That topic was chosen because we are all familiar with and frustrated by the difficulties dealing with shopping carts in crowded supermarkets. The ubiquity of the product, which makes everyone an "expert," and the near-universal negative experience with it would seem to make it a challenging case. To make things even more difficult, IDEO agreed to produce a new shopping cart in a week. The result was remarkable; in contrast to the heavy and bulky shopping cart we are all familiar with, IDEO produced a sleek modernistic machine, with wheels that rotated to allow maneuvering in narrow supermarket aisles (see Column I in Table 2–2). The large metal basket was replaced by several smaller plastic ones that were removable. The child's seat was equipped with a safety bar as well as a play surface. The cart had a microphone to allow communication with customer service in the supermarket and a scanner that allowed self-checkout. Finally, the frame of the cart contained hooks to hold the bags of purchased goods after the baskets had been removed while checking out.

We can now place those advances in the context of the questions asked so far in this chapter. Although the new shopping cart is impressive as an accomplishment, should we conclude that all or most of its components

Table 2–2 Novel components of shopping cart and where they came from.

I. NOVEL COMPONENT	II. DIFFICULTY	III. BASIS FOR INNOVATION
1) Casters allow sideways movement	Hard to navigate in aisles	**Near analogy:** wheels and casters
2) Plastic basket	Pilferage: Metal baskets serve as barbecues	**Logic**
3) Small baskets can be removed and carried	Hard to navigate: Easier to use cart as "home base" and bring items to cart	**Logic**
4) Safety bar on child's seat	Child Safety: Unattended child leaves safety seat	**Near analogy:** "safety seat" on roller coaster
5) Play surface on child's seat in cart	Child irritability	**Logic:** play reduces child's irritability
6) Microphone in cart	Difficulty finding items ⇒ Need to contact customer service	**Logic** + **Near analogy** (cell phone?) + IDEO's "electronic gadgets" **expertise**
7) Scanner in cart	Checkout can be slow: Avoid lines ⇒ Self-checkout	**Logic** + IDEO's "cyberize" **expertise**
8) Hooks for bags on frame after baskets are removed checking out	Transporting heavy bags to car	**Logic**
9) Sleek design	Ugly shape	**Expertise:** Designers' sensibilities

came about through out-of-the-box thinking? That is, did IDEO break with the past in coming up with the new? I contend that IDEO's advances were no different than those discussed earlier in this chapter: the components of the new cart can be seen as building on what had been known and going beyond it through ordinary thinking.

When IDEO accepted Koppel's challenge, the first step they took was to investigate the domain they were to work in (see Figure 2–3): they observed professional shoppers in supermarkets (people who went through supermarkets in order to purchase items for internet buying services) as well as ordinary people shopping in supermarkets. IDEO also contacted an individual who bought shopping carts for a supermarket chain. It became clear to the IDEO staff that there were various problematic aspects of shopping carts (see Column II in Table 2–2). We can now see that those novel components of the shopping cart were responses to problems and difficulties found in the cart. The critical question to be examined is how each of those novel components came about: What can we say about the thought processes underlying each? More specifically, did out-of-the-box thinking underlies those innovations?

The answer to that question seems to be no: For all the shopping-cart problems shown in Table 2–2—for all the novel ideas generated—the same

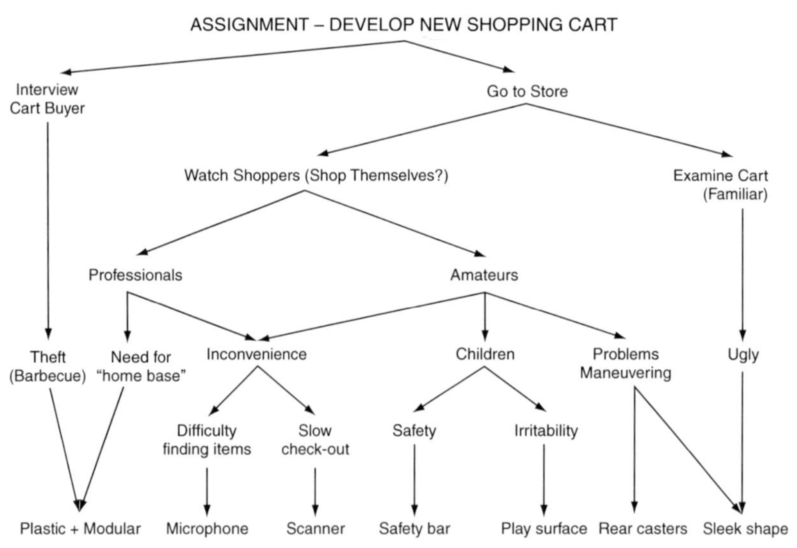

Figure 2–3 The paths to a new shopping cart.

processes were used as seen in the case studies discussed earlier. First, we see analogical transfer, the transfer of knowledge from one situation to an analogous situation. As with the other case studies, the examples of transfer found in the shopping cart were based on near analogies (see #1, 4, and 6 in Column III in Table 2–2). As one example, the use of casters to allow movement in the supermarket isles (#1) was taken from office chairs. Similarly, the safety bar on the child's safety seat was taken from seats on rides at amusement parks, where it serves the same purpose; that is, to keep people safely in their seats. Second, we also see the use of logic as the basis for deducing the solution to a problem (see #2, 3, and 5–8 in Table 2–2). This is clearly seen in the use of plastic baskets: the cart-buyer noted that there was a large amount of pilferage of shopping carts from supermarkets, because the carts' large metal baskets were useful as barbecues. This statement of the problem leads almost directly to the solution: if people are stealing carts because they have a large metal basket, then make the basket non-metal.

Similarly, observation of the professional shoppers (#2) led to the realization that those people made frequent trips away from the stationary cart, since it was easier to leave the cart and walk or run to get the various items. This led to equipping the cart with small baskets that could be removed and carried, to assist the shopper in accumulating larger numbers of items on each expedition away from it. Finally, the expertise of the IDEO staffers played at least a partial role in bringing about some of the innovations in the shopping cart (see #6, 7, and 9 in Table 2–2). The presence of the microphone for contacting customer service, the built-in scanner, and the sleek overall shape of the cart were the results of the expertise of the IDEOers concerning electronic gadgets and as designers.

The case study of IDEO's development of a new shopping cart has produced results that parallel those found in the other case studies. We found little evidence for out-of-the-box thinking, and did find support for the idea that creative thinking is the result of the same processes that underlie ordinary thinking. The IDEO group were attempting to solve a set of problems that they developed from their observations of users of the shopping cart, and they developed solutions to those problems by (1) using their expertise as designers, (2) transferring information from situations analogous to those they were facing, and (3) using logic to draw conclusions that resulted in novel responses to the situation they were facing.

General Conclusions from Case Studies

The most important conclusion drawn from this set of case studies is that creativity can indeed be based on "thinking inside the box." The creative advances just discussed, which came from a reasonably broad sample of domains, did not come about through the rejection of the past. In all the cases we examined, the past served as the basis for the construction of the new. Furthermore, the thought processes used in that construction are "ordinary" thought processes that are seen in all our interactions with the world. We have seen the use of domain-specific expertise, based on extensive training within a domain. We have also seen examples of transfer based on analogies. Finally, we have seen examples of creative thinking based on logical reasoning. Those findings can be seen as having implications for our understanding of how one might foster innovation in industry on a broad scale.

On Fostering Innovation

One general implication of potentially great importance to be derived from the case studies discussed in this chapter is that we are all capable of creative thinking. If we can conclude from the case studies that the processes involved in creative thinking at its highest levels are those that we all carry out in our ordinary activities, then, *ipso facto*, we are all capable of thinking creatively. Furthermore, it may be true that we carry out creative thinking all the time, without thinking about it (Weisberg, 2006). If we consider the fact that our world is constantly changing, which requires that we must constantly adapt our behavior to novel circumstances, then it follows that we are always being creative, at least on a small scale. Although our day-to-day creativity might not be on the level of the advances discussed here, the thought processes might still be essentially the same. A further important issue that arises from that conclusion is that any differences between the "creative geniuses" and the rest of us are not based on thinking processes. If the genius does not make far-ranging creative leaps that we ordinary folks are incapable of, then any differences in creative accomplishments among us must not be due to basic differences in thinking processes. This of course leads to the question of what the differences between the geniuses and ordinary people might entail, and one variable might be motivation. That is, the seminally creative individual might be willing to work harder than other people do, which would mean that he or she would acquire a deeper and richer database than others do. In addition, those individuals might work harder in order to apply the information from that database to the problems that they face.

How does one find opportunities for innovation—i.e., problems to solve?

Consider the problems discovered with the shopping cart: IDEO went to shoppers in order to learn from the "front lines" exactly what were the basic problems in using the shopping cart. The types of observations made by the IDEO staff were of two sorts. First, they observed ordinary shoppers, from whom they learned about the problems facing typical shoppers in their day-to-day interactions with the shopping cart; every shopper could have produced that list and more. Second, they also observed professional shoppers as they did their shopping. The difference between professional and ordinary shoppers is that the former are working under much stronger time pressures than the latter. Those extreme time pressures probably resulted in the professional shoppers' developing strategies for dealing with shopping—and with shopping carts—that went beyond those developed by ordinary shoppers, but which nonetheless would be useful to many of us (see also von Hippel, 2005).

So here is the first step in finding opportunities for innovation: Do what the users do. Product-development personnel often contact consumers and interview them (or carry out focus group interviews) in order to determine what difficulties they see in a product, so that the product can be improved. However, people cannot always describe the difficulties they experience using a product, so there may be much potentially important information that is not obtained. The analysis of IDEO's development of the new shopping cart points out how one can best obtain information concerning how a product might be improved: Don't *ask* people how they use a product and what problems they have with it: *observe* them or *become* them, so that all the potential problems can be seen first-hand. As an example, if you run a retail company that does online business, go to the web site and try to order something; call customer service and try to have a problem straightened out. If you run a company that creates financial software, use your program to prepare your taxes. If you head a company that makes razors, try to use the razor to shave. If you head a company that makes a stain-removing product, try to use the product to remove the stain. Also, find out from consumers themselves how they use your products. Certain select consumers—called "lead users"—often find difficulties with products and try themselves to overcome those difficulties (von Hippel, 2005). Those problems and their solutions can be the basis for new innovations. We saw some examples of "lead user" innovation in the analysis of the problems with the shopping cart.

Once a problem—an opportunity for innovation—is located, how should the would-be innovator approach it? As we have seen, some problems are solvable through the use of general knowledge: logical thinking and related skills that we have all acquired over our lifetimes. Here is a problem: people steal shopping carts because they contain a large metal basket. That leads to the solution of making the basket non-metal. All that was needed for the problem to be solved was that the problem be made explicit. Other problems require expertise. If the problem is in your area of expertise, then use that expertise—don't look for wild leaps out of the box. A problem typically will not be identical to something you have seen before (otherwise it would not be a problem). Therefore, you may have to work to find a match between what you know and the difficulty you are facing. Examine the problem from a more general perspective, so you go beyond identity. Here is an example: you want to improve the maneuverability of the shopping cart. Go beyond shopping carts to other movable objects with wheels. You can think of the shopping cart as an object to be moved with wheels. Do you know other objects of that sort whose design might be useful in redesigning the shopping cart? Office furniture moves on wheels; perhaps the casters on office furniture could serve on a shopping cart. Thus, one solves this problem through an analogy, and the strategy of looking for analogies can be broadly useful in solving problems, including those involving innovation in a corporation. Do you know anything that you can apply to the problem?

This discussion can now come back to an issue discussed at the beginning of the chapter: the use of idea factories; that is, the question of solving problems in innovation in-house versus outsourcing. Should you seek novel ideas from "idea factories"? Outsourcing innovation to idea factories is based on two assumptions. The first is that domain-specific expertise is not critical in producing new ideas. (If expertise were critical, then why should you go beyond your in-house experts?) One goes outside for help in innovating if one assumes that novel ideas come about through out-of-the-box thinking—a set of general "creativity" skills that you and your employees do not have. The idea factory can use those general out-of-the-box thinking skills to produce the ideas that your people cannot produce. The second assumption is that the idea-factory people will not get caught up in the constraints brought about by expertise.

The case studies indicate that both of those assumptions are incorrect. First, we have seen that knowledge and expertise are critical in development of new ideas. Out-of-the-box thinking is not the basis for new ideas. Therefore, innovation should be done in-house. Furthermore, the specific examination of the case of IDEO's creation of the new shopping cart provides evidence directly related to the question of whether one should attempt to outsource problems involving innovation. The new shopping cart was the result of processes no different than those involved in ordinary thinking. IDEO did not do anything on a cognitive level that could not have been done in-house by a shopping-cart manufacturer with designers and engineers on its staff. Should a company ever expect to outsource innovation? Based on the cases discussed in this chapter, yes. A company might want to go beyond in-house experts when the outside source has expertise concerning the type of problem at hand. The first thing that IDEO's staffers did was to obtain as much information as possible about problems with shopping carts, by tapping into the experiences of experts and using those experiences as a surrogate for expertise of their own. IDEO's staff quickly became much more knowledgeable about the problems with shopping carts than most of us are, and this knowledge resulted in the development of a set of problems to which they could apply the expertise that they brought to the problem.

A further possible difference between IDEO and the typical R&D department of a corporation may be that the IDEO corporate culture is steeped in innovation (Sutton and Hargadon 1996). That is, IDEO is staffed by engineers and industrial designers whose careers have been based on their ability to produce innovations. Furthermore, when a potential client comes to IDEO with a problem, that problem is dealt with by a group of individuals of wide-ranging expertise, each of which is relevant to the problem in a different way. As an example, when a manufacturer of goggles came to IDEO to ask for their help in designing a new goggle, people who were asked to join the group working on the project included one person with expertise in clear plastic, another with expertise in foam, and another with expertise in manufacturing. IDEO brings together people who comprise a *super-expert* to deal with a problem. In addition, when a new problem comes to IDEO, the client spends time explaining the difficulty to the engineers and designers who will be working on it. Those individuals also spend time preparing for the project by reading as much as they can about the product and also by examining the client's current product, if any, as well as those of the competitors. Thus, one could say that the staff at IDEO undergo an immersion in the client's problem so that they can become experts in that specific area, and so that they can then extend their existing expertise to the new problem.

Final Thoughts

The discussion in this chapter has indicated that creativity and innovation are within everyone's capabilities. The thought processes underlying innovation are those that we use all the time in our professional lives and our personal lives. It must be emphasized, however, that creativity and innovation do not come easily. First, you who wish to innovate must first work to find new problems. Second, you must also work to broaden and deepen your database in order to solve them.

References

Chipp, H. B. (1988). *Picasso's "Guernica": History, transformations, meanings.* Berkeley, Calif.: University of California Press.

Frensch, P. A., and Sternberg, R. J. (1989). Expertise and intelligent thinking: When is it worse to know better? In R. J. Sternberg (Ed.), *Advances in the psychology of human intelligence,* Vol. 5 (pp. 157–188). Hillside, N.J.: Erlbaum.

Heppenheimer, T. A. (2003). *First flight: The Wright brothers and the invention of the airplane.* Hoboken, N.J.: Wiley.

Judson, H. F. (1979). *The eighth day of creation: Makers of the revolution in biology.* New York: Simon and Schuster.

Kelley, T. (2000). *The art of innovation: Lessons in creativity from IDEO, America's leading design firm.* New York: Doubleday.

Olby, R. (1994). *The path to the double helix: The discovery of DNA.* New York: Dover.

Osborn, A. (1953). *Applied imagination.* Revised edition. New York: Charles Scribner's Sons.

Runco, M., and Pritzker, S. (1999). *Encyclopedia of creativity.* San Diego, Calif.: Academic Press.

Simonton, D. K. (1995). Foresight in insight? A Darwinian answer. In R. J. Sternberg and J. E. Davidson (Eds.), *The nature of insight* (pp. 465–494). Cambridge, Mass.: Massachusetts Institute of Technology Press.

Simonton, D. K. (2003). Scientific creativity as constrained stochastic behavior: The integration of product, person, and process perspectives. *Psychological Bulletin, 129,* 475–494.

Sternberg, R. J. (1999). *Handbook of creativity.* New York: Cambridge University Press.

SUTTON, R. I., and HARGADON, A. (1996). Brainstorming groups in context: Effectiveness in a product design firm. *Administrative Science Quarterly, 41*(4), 685–718.

VON HIPPEL, E. (2005). *Democratizing innovation.* Cambridge, Mass.: MIT Press.

WEISBERG, R. W. (2004). On structure in the creative process: A quantitative case-study of the creation of Picasso's *Guernica. Empirical Studies in the Arts, 22,* 23–54.

WEISBERG, R. W. (2006). *Creativity: Understanding innovation in problem solving, science, invention, and the arts.* Hoboken, N.J.: John Wiley.

WEISBERG, R. W., and HASS, R. (2007). We are all partly right: Comment on Simonton. *Creativity Research Journal, 19,* 345–360.

CHAPTER 3

"PUTTING BLINKERS ON A BLIND MAN"

PROVIDING COGNITIVE SUPPORT FOR CREATIVE PROCESSES WITH ENVIRONMENTAL CUES

BO T. CHRISTENSEN

CHRISTIAN D. SCHUNN

ARE you stuck on a creative problem, and don't know where to go from here? Try this: In what ways might you use a stork to solve your problem? A key chain? A foreign country? Two friends? A pair of pliers? Have you solved the problem yet?

Random or blind input into the ideational stages has long been thought to be potentially beneficial for solving creative problems. Theoretically, this position was forcefully put forth by Campbell, who in his 1960 article argued that "a blind-variation-and-selective-retention process is fundamental to all inductive achievements, to all genuine increases in knowledge, to all increases in fit of system to environment" (Campbell, 1960, p. 380). In other words, in taking a blind dip into the infinite sea of ideas, you are better off generating more ideas than fewer, since it's not possible to know up front which ideas are going make it into history rather than the recycle bin (see also Simonton, 2003). In that sense, producing more ideas will help your chances of getting a good idea, and input that helps you associate or relate to distant or novel areas in your thinking processes may help you along in getting more ideas. Several approaches have been developed to help the creative problem-solver produce more ideas, the most well-known being brainstorming (Osborn, 1963). Brainstorming essentially uses a group setting for striving for a multitude of novel ideas, where each idea acts as essentially random input to further idea-generation by the other group participants. Other practices that have attempted to use random stimuli to promote creativity include "forced connections" (e.g., Firestien, 1996; MacCrimmon and Wagner, 1994; Terninko, Zusman, and Zlotin, 1998), where the problem solver attempts to relate to a random picture or other complex stimuli, and use that input in problem solving. De Bono (1975) developed the dictionary method, which is simply to introduce a random word by, say, selecting a random page and position of the word in the dictionary, and then use that word to solve your problem (see de Bono, 1992, for more methods involving random input). MacCrimmon and Wagner (1994) developed and tested software for support for a number of these random input techniques, and concluded that: "We feel confident in asserting that 'forced connections work' but have no detailed evidence on which ones work better under particular circumstances and, more importantly, cannot explain why they work" (p. 1531).

Some potential explanations for how forced connections might work have been put forth in the cognitive sciences. Notably, the literature on analogical transfer (Forbus, Gentner, and Law, 1994; Gick and Holyoak, 1980; Holyoak and Thagard, 1995) has dealt extensively with trying to understand how subjects retrieve and transfer elements from past examples to new problems. Using Duncker's (1945) radiation problem, Gick and Holyoak's classic study (1980; 1983) demonstrated how subjects transferred elements from the Attack-Dispersion problem to help solve the radiation problem. The study also showed that, unless explicitly instructed to try to make a connection between the problems, subjects rarely noticed their similarity and failed to use it to solve the radiation problem. This observation of the lack of an automatic transfer has been repeated many times since (see, e.g., Anolli, Antonietti, Crisafulli, and Cantoia, 2001). Some of the reasons for the lack of automatic transfer have been found to include the difference between superficial and structural similarity (Gentner, Rattermann, and Forbus, 1993; Holyoak and Koh, 1987), in that it has become evident that whereas structural similarities may ensure effective transfer, the superficial similarity between concepts is a powerful driver of noticing the connection between them in the first place. Without superficial similarity, spontaneous retrieval is rare. Furthermore, if people are instructed to actively look for connections between concepts or domains, they produce far more transfers than if no instructions are provided.

Associative theory provides another potential explanation for why forced connections with random input might improve creativity. Mednick (1962) was among the first to propose an associative theory of the creative process, in asserting that creative ideas tend to be *remote*. That is, original ideas usually only come to you once the obvious ones are depleted. For example, in rating open-ended responses to questions such as "name all of the uses of a brick you can think of," the second half of ideas will tend to be more original responses than the first half (Mednick, 1962; Runco, 1985).

Empirical studies have tried to establish that exposures to *relevant* cues lead to increased performance on creative tasks, but they have shown mixed results for both analogies and associations (see Christensen and Schunn, 2005, for a brief review), with some studies showing positive effects, and other studies showing null results. Empirical studies trying to establish that *random* stimuli lead to increased performance, however, seem to be almost nonexistent. Insofar as random cues will tend to include at least some elements that can potentially help solve creative problems, the mixed findings from the relevant cuing conditions may generalize to random conditions.

So, all in all, there appears to be some (although mixed) evidence that providing exposures to random cues can increase performance on creative problems, based on theories of analogical transfer and association. Many a practitioner may settle for this as evidence enough—"Random stimuli may enhance creativity, so let's try it—what harm could that do?" Well, considering another classic line of creativity research, potentially quite a bit!

Gestalt and cognitive psychologists have long been interested in the potentially harmful effects on creativity and problem solving of past knowledge, strategies, and behavioral tendencies. For example, Wertheimer (1959) used the term *reproductive thinking* to refer to problem solving in which one blindly carries out a previously learned procedure. Early gestalt studies by Maier (1931) and Duncker (1945) showed how the standard functions of objects apparently restricted individuals' ability to use them in novel ways in creative problem solving, a phenomenon termed *functional fixedness*. Luchins (1942) went on to demonstrate that having problem-solving strategies may lead people to become unable to solve new problems that do not lend themselves to being solved with the standard strategy, despite the fact that the problems could be easily solved if no problem-solving strategy had been established. This phenomenon was termed a "mental set." Cognitive studies of fixation have looked along similar lines, showing how people can become fixated on certain solutions, leaving them unable to solve creative problems, or solving them less creatively than without such fixating elements. Such fixation may occur if a person is introduced to an element in the environment that will tend to lead his thinking in certain directions; for example, by trapping his thinking in a "mental rut" through providing enticing and well-known (but unsuitable) solution elements (e.g., Smith, 1995a, 1995b), similar to the experience of knowing the answer but being unable to currently produce it (a.k.a. the Tip-of-the-Tongue phenomenon).

A number of studies have shown how providing (Jansson and Smith, 1991; Ward, 1994; Dahl and Moreau, 2002; Marsh, Ward, and Landau, 1999; Jaarsveld and van Leeuwen, 2005) or retrieving (Ward, 1994) existing examples may inhibit generative creative processes. Examples, in this sense, lead to a higher proportion of property transfers from the examples into the subject's own work (e.g., Marsh, Landau, and Hicks, 1996), and notably this result also occurs even when subjects are explicitly instructed that they should try to avoid such transfer (e.g., Smith, Ward, and Schumacher, 1993). Extending these findings, it has been shown that, especially in generative tasks, people are frequently incapable of monitoring this property transfer (e.g., Marsh, Landau, and Hicks, 1997). In such cases, the unconscious influence of memory causes current thoughts to be (wrongly) experienced as novel or original inventions, which is also termed "unconscious plagiarism" or "cryptomnesia" (Brown and Murphy, 1989; Marsh and Bower, 1993; Marsh et al., 1999; Marsh and Landau, 1995). Ward (1994, 1995, 1998) proposed a path-of-least-resistance model to account for some of these findings, which states that the default approach in tasks of imagination (especially when few constraints must be satisfied) is to access a specific known entity or category exemplar (gravitating towards basic level), and then pattern the new entity after it. In support of this model, Ward (1994; Ward et al., 2002) found that people who report basing their novel constructions on specific exemplars are less original than people who use other strategies. Property transfer in generative tasks has proven robust across a variety of settings, including engineering design tasks conducted in the lab (Jansson et al., 1991; Dahl et al., 2002; Christiaans and Andel, 1993). Jansson and Smith (1991) had people (either mechanical engineering students or professional designers) work on simple design problems (such as how to construct a car-mounted bicycle rack), with (the fixation group) or without (the no fixation group) a specific example provided by the experimenter. They found that the fixation group included more properties from the examples. However, it should be noted that a failure to replicate this finding has been reported (Purcell and Gero, 1992).

Reproductive-thinking theories such as these are basically saying that if you use your past behavior, strategies, and representations to solve novel or creative problems, then you may end up with a less-than-original solution. The theory aligns well with the frequently used search metaphor in creative problem solving, by showing how searching for creative solutions using past knowledge may lead you down the wrong path, into mental ruts (Smith, 1995a, 1995b), or to an oasis of false promise that is hard to leave again (Perkins, 2000). Reproductive theories contrast with the classic problem-solving literature, which has listed the use of past knowledge and heuristics as being useful to problem solving (e.g., Newell and Simon, 1972; see also Metcalfe and Wiebe, 1987). It seems that specifically in creativity, past strategies, exemplars, or knowledge may in some cases (but not all: see Ward, Smith, and Vaid, 1997) be harmful.

So, this may all be quite confusing to the practitioner interested in using random stimuli. Will random stimuli enhance creativity—or does he risk jeopardizing the creativity of the outcome instead? A first step in resolving this dilemma obviously involves understanding that not all environmental cues are beneficial to a particular creative task—the environment is not always kind—and some environmental cues will have quite the opposite effect; they will be fixating or detrimental to problem solving. But which ones? It is not enough for the practitioner to "avoid the fixating stimuli," since that would be impossible to know *a priori*. Put simply, part of the solution may lie in knowing whether the problem solver is expected to use a stimulus as a cue for past reproductive behavior, or as a cue to generate new solution attempts. But it is still more complex than that: creativity is not a single process, but a host of processes, and pooling all creative processes provides an insufficient understanding of what creativity is about. At a general level, creative processes can be divided into generative and exploratory processes (Finke, Ward, and Smith, 1992). Generative processes are, for example, analogical transfer, association, retrieval, and synthesis; while the exploratory processes include, for example, contextual shifting, functional inference, and hypothesis testing.

So not all cognitive processes involved in creativity are generative processes. A random cue may, for example, promote one type of generative creative process, while it limits another exploratory one. Cuing very distant domains may help your mind wander widely, but it is likely to distract attention from closely scrutinizing and testing the current idea. It is necessary to try to predict whether a cue may help or hinder, depending on the expected function of the creative cognitive process. To this end, a more detailed understanding of the individual creative processes and their functions is needed.

If we are to help the practitioner in his or her creative process, it is necessary to dive deeper into the processes and functions in creativity, in order to further understand how and why some stimuli may start up creativity at certain points, while stopping it at others. Here we will limit ourselves to examining two creative processes: analogical transfer and mental simulation. By understanding how these processes work and which functions they serve, we aim to provide some guidelines for the practitioners in selecting and using somewhat random or blind stimuli to promote creative output. Rather than just looking at the connection between cue and creative outcome, we instead examine the intermediate factors of how environmental cues affect creative cognitive processes, showing that the cue effect depends on the functions served by the cognitive processes. And further we examine the link between the creative cognitive processes and the creative products resulting therefrom. The hypothesis is that a strategic interaction between creative cognitive processes and environmental stimuli can lead to products that are more original, useful, and thus creative (cf. the standard definition of creativity, Mayer, 1999).

We have studied both of these processes in real-world engineering design, using the *in vivo* method (Dunbar and Blanchette, 2001). Basically, *in vivo* research entails selecting a suitable object of study in the real world, collecting data from these objects of study using video or audio recordings, transcribing the recorded data, and coding the data according to suitable coding schemes. Dunbar (1995, 1997, 2001b) has used this methodology to examine analogy in the domain of science (in particular, microbiology), where he found that a suitable object of study for the study of scientific thinking and reasoning was the weekly laboratory meeting. *In vivo* methodology has the advantage of adding ecological validity to the typical laboratory experiments conducted in cognitive science, at the expense of having limited control over individual variables. As such, it is well suited for exploring functions and mechanisms of real-world creative processes that can then later be examined further in laboratory studies. We used the methodology to examine functions of creative cognitive processes in engineering design conversations.

A number of terms will be used here to denote the objects presented during the creative process and their impact on subjects' thinking. "*Prime*" is used to denote objects where the subject is not consciously aware of the impact of the object on his or her thinking or outcome. "*Input/stimuli*" is used as a neutral marker of the objects presented to the subject during creative processes, regardless of whether the subject uses this object in thinking or creative solution. "*Cues*" usually have positive connotations in that, in order for something to be a cue, a third party (usually the experimenter) is assuming that the object should make a positive contribution to problem solving. However, here "*cue*" is used in the sense "blind or random cue," since here it is not known beforehand whether a given "cue" object will have a positive or negative impact on creative problem solving. The term "*relevant cue*" will be used in those instances where a third party has evaluated *a priori* that the cue ought to make a positive contribution to problem solving, and the term "*fixating cue*" will be used where a third party has determined that the cue will have a negative impact.

Analogical Transfer and the Relation to Random Cues

Analogy involves accessing and transferring elements from familiar categories (source) to use in constructing a novel idea; for example, in an attempt to solve a problem or explain a concept (target) (Gentner, 1998). In design as in other creative domains, analogy has been argued to be of special importance (e.g., Roozenburg and Eekels, 1996; Casakin and Goldschmidt, 1999; Goldschmidt, 2001), as also evidenced by the many anecdotes of breakthrough inventions following distant analogies that exist in the design field. One of the most famous anecdotes is George de Mestral's development of Velcro® after examining the seeds of the burdock root that had attached themselves to his dog. The sheer number of similar anecdotes of breakthroughs and inventions attests to the importance that is placed on analogy in domains of innovation (see, e.g., Shepard, 1978; Ghiselin, 1954). Below we will look at analogical functions, analogical distance, and the automatic nature of analogical retrieval in order to show how they relate to random stimuli influx.

Analogies Serve Other Functions Besides Problem Solving

Analogies are constructed for different purposes. While it is the problem solving or generative functions of analogies that have interested creativity scholars and cognitive scientists, analogies have been shown to have other functions in science and design. Notably, Dunbar's (1997, 2001a) *in vivo* studies of real-world analogizing in science distinguished four types of functions for analogies: forming hypotheses, designing experiments, fixing experiments, and explaining concepts to other scientists (see also Ward, 1998, for another classification of analogies in invention). Dunbar found that almost half of the analogies were explanatory.

These functions are, however, in part specific to science. In design, for example, other kinds of activity are more prevalent and important, such as the construction, modification, and evaluation of novel and useful objects. In our own real-world studies of engineering design (Christensen and Schunn, 2007), we found that analogies served three different functions: explanation, problem solving, and problem identification. The first two are similar to analogy functions in science, while the last one has been uniquely identified in design. Engineering design is frequently conducted in teams, rather than individually, whereby communicating novel ideas to other members of a team becomes an important part of the process. *Explanations* through analogy can be a way of enhancing and ensuring comprehension while avoiding misunderstanding when dealing with novelty. Particularly when novel ideas are discussed that are unsupported by diagrams or prototypes, explanatory analogies serve an important purpose, as it is can be difficult for design team members to know whether they are referring to the same thing. Here analogies serve the purpose of communicative alignment in design conversations.

Yet another function analogy serves is that of *problem solving*. Indeed as noted above, this function is perhaps the primary reason researchers have focused on analogy in design and science. In addition to these two functions, *problem identification* is an important function, especially in the early conceptual stages of engineering design. When developing novel concepts, it is necessary to try to foresee whether a novel idea or concept would work under particular circumstances. In this case, analogy plays a part in evaluating novel concepts, in that it is possible to transfer, not only solutions but also potential problems from sources with which the designer has past experience. Here the elements to be transferred from source to target involve potential design problems that the new concept may display. In engineering design, we found that the functions of analogies were distributed roughly evenly among these three categories, with 32 percent explanatory analogies, 40 percent problem-solving analogies, and 28 percent problem-identifying analogies.

These real-world findings lend support to the hypothesis that analogies do not serve a single purpose in science or design. Rather, in design, it seems that analogies are used for widely different cognitive functions, such as explanatory communicative support (e.g., using an analogy to a concept known by all participants in order to promote comprehension of a new and unfamiliar concept), generative processes such as problem solving (e.g., transferring a structure from an exemplar into creating new solutions), and exploratory processes such as problem identification (e.g., transferring problems with a previous structure to a new structure in order to identify problems with the new structure). Since analogies do not serve a single function in design, it is necessary to identify which analogy function you wish to promote, before going to some lengths in order to promote the use of analogies. For example, promoting the use of analogies may not help you generate new ideas if the analogies are merely used for explanatory purposes. Furthermore, this analogy function distinction suggests that analogies can be used in other aspects of creative work than to generate solutions and solve problems. In exploratory stages, where new ideas need to be closely examined, tested, and scrutinized, promoting analogy use may lead to identifying more design problems based on previous experience.

Within-Domain Sources Are Easier to Retrieve But May Lead to Less-Than-Original Responses

In analogical transfer, the "distance" between source and target may be large or small. To cite examples from our own research, a designer trying to develop a new type of blood bag in medical plastics may, for example, make an analogy to other blood bags in medical plastics (within-domain, or local analogies), or make an analogy to Christmas decorations or shoes or credit cards in developing the design (between-domain, or distant analogies) (Christensen and Schunn, 2007; see also Dunbar, 1995; Dunbar et al., 2001; Vosniadou and Ortony, 1989). A consistent finding in the research literature is that transfer increases with similarity (e.g., Holyoak et al., 1987; Novick, 1988; Ross, 1987; 1989; Simon and Hayes, 1976). But whereas analogical transfer has been found to be closely related to structural similarity, analogical access often strongly depends on superficial similarity between source and target (Gentner et al., 1993; Holyoak et al., 1987; Novick, 1988; Ross, 1987). The distinction between local and distant analogies is related to the differential amount of superficial similarity, with more superficial similarity for local analogies. This higher amount of superficial similarity may make local analogies easier to access (e.g., Gentner et al., 1993; Holyoak et al., 1987). Furthermore, both local and distant analogies contain structural similarity, but since distant analogies connect two previously distinct concepts or domains, it may be more difficult to ensure a successful transfer of solution elements in design problem solving from source to target because the domains may differ in multiple subtle but important ways (Johnson-Laird, 1989).

Few studies have looked at the use of distant analogies in design. In an experimental study of visual analogy in design, Casakin (2004) found that both novices and experts produced more between-domain than within-domain analogies. While this study and the above-mentioned design anecdotes suggest that between-domain analogizing may be common in design and science, naturalistic studies of analogy in science seem to question this conclusion. Dunbar (1995; 2001a) found that distant analogies did not play a significant part in discovery, but rather were very rare in comparison to local analogies. However, our real-world research in engineering design has shown that in design local and distant analogies are about equally prevalent (Christensen and Schunn, 2007), indicating both that between-domain analogizing is common, and that within-domain analogizing is also used heavily in design.

The research on fixation and exemplar influence in generative tasks described above supports the notion that having or making examples available will bias people's creations toward features in those examples. Objects from similar domains share more superficial similarity than objects from dissimilar domains, and since superficial similarity is one of the key driving forces of analogical access, this leads to the expectation that the presence or availability of within-domain exemplars increases the likelihood of within-domain analogizing (Ward, 1998). In other words, the presence of within-domain examples may make it hard for creative problem-solvers to break away from local analogies, since superficial similarity dominates access, and distant analogies will be less superficially similar than local analogies. Providing prior within-domain examples thus biases people's creations toward features contained in those examples (Marsh et al., 1996). This within-domain biasing could be the case, for example, when designers use external support of prototypes during the concept phases in engineering design, as compared to conditions without such support. We examined this issue in our real-world engineering design study (Christensen and Schunn, 2007), by comparing analogy frequency with or without the reference to sketches and prototypes. Here it was found that the prevalence of between-domain analogies in design conversations was reduced when referring to within-domain prototypes as compared to unsupported cognition. This result suggests that, in real-world design, the use of within-domain tools may in effect be limiting the use of generative between-domain analogies, thus extending the unconscious-plagiarism-finding that making within-domain exemplars available during the creative constructive process tends to lead subjects to unconsciously plagiarize these exemplars, to include objects that are a natural part of the design space. If exemplars are present, the designers are less likely to think about other domains than the present one.

These findings can be substantiated by Dahl and Moreau's (2002) study of analogy use in design. They had undergraduate engineering students design new products that would solve problems for the commuting diner (e.g., difficulties with spillage, consumption, and storage of food during automotive driving). They found that exposing students to one or several within-domain examples (e.g., sketches of a drive-in window) led to a lower proportion of distant analogies being used compared to subjects who saw no sketch. Furthermore, the proportion of distant analogies used was a strong indicator of the originality of the resulting design. Apparently, the presence of one or more within-domain exemplars hindered students in producing original responses.

More tentative support for the link between external within-domain sources to analogical use and outcome comes from experiments providing visual analogs as hints in problem solving (Beveridge and Parkins, 1987) and design (Bonnardel and Marmèche, 2004; Casakin et al., 1999). These experiments indicate that providing visual information can lead to transfer of solution elements. These findings could have important implications for structuring innovation tools. If the designer's goal involves generating novel and original products seemingly unrelated to or inspired by previous products in the category, then a tentative recommendation would be to avoid using environmental cues that point towards within-domain exemplars. Apparently the tendency to access and transfer within-domain exemplars into novel work is quite potent, and even extends into the presence of typical design-space objects (such as prototypes) not intended to lead design thinking along traditional (as opposed to original) paths.

We examined *in vivo* whether analogical distance would interact in certain ways with analogical functions. Based on past research, our expectation was that explanatory analogies would be primarily between-domain, since between-domain analogizing may be necessary in explaining novel design concepts precisely because they are new to the domain. Problem identification in evaluation was hypothesized to involve primarily within-domain analogies since within-domain analogies are more accessible due to superficial similarity, more available due to within-domain expertise, and more appropriate for identifying problems because within-domain analogies may increase the chances of successful transfer. And finally, because engineering design involves the production of novel and useful solutions, solving problems by relating to past within-domain knowledge may frequently not be enough to construct an original product. Therefore, a mixture of within- and between-domain analogies is expected when the function of the analogy is to solve a design problem. As expected, these three types of analogy functions in design had differential ratios of within- to between-domain analogies. Problem identification analogies were mainly within-domain, explanatory analogies were more frequently (and mainly) between-domain, while problem-solving analogies involved a mixture of within- and between-domain analogies (see Fig. 3–1).

These findings on analogical distance have a number of consequences for tools supporting the use of analogies in innovation. As noted under analogical functions, when promoting the use of analogies it is important to take into consideration which analogy function to support. If the purpose is to support the problem-solving or generative aspects of analogies, then both within-domain and between-domain products may be used. However, insofar as local analogies are used, it can be expected that a fair amount of exemplar property transfer will take place, thereby making the resulting innovation structurally similar to previous exemplars. This similarity may in effect reduce the evaluated originality of the resulting product. This does not, however, mean that within-domain exemplars are useless in solving problems. Quite the contrary—within-domain analogies may be quite effective in solving problems in design. In fact, it is quite possible that within-domain analogies may be quite effective in making the resulting design solutions more useful, since the sources may be well-known sources or industry standards that are effective in securing transfer of proven and comprehensible solutions. But the resulting innovation may not be very original, so within-domain analogizing should perhaps primarily be used in

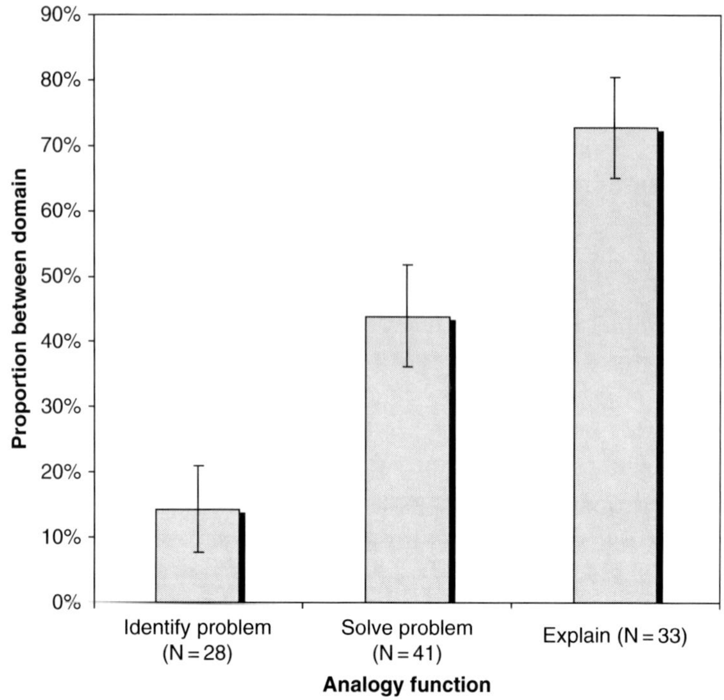

Figure 3–1 Proportion between domain analogies by analogical function (with SE bars).

cases where the resulting design is not required to be original or notably distinct or different from previous designs. So, while within-domain analogizing is effective at solving problems, property transfer from within-domain exemplars will tend to make the resulting solution less than original.

Furthermore, as several theories have pointed out, within-domain analogies (sharing a greater amount of superficial similarity) will tend to be retrieved and used more easily. Therefore, it could be argued that within-domain analogies will be retrieved and used more frequently, be considered more relevant and interesting, and examined more closely, when compared to between-domain sources. This tendency should tend to lead to within-domain exemplars' overshadowing between-domain exemplars, insofar as they are both used as cues. A recent study tested this prediction and showed that within-domain exemplars drew more attention (were looked at longer) and were considered to be much more inspiring to the designers, compared with between-domain exemplars or other cue categories (Christensen, under review).

Additionally, analogy serves other functions than problem solving. It may be possible to promote in particular the problem-identifying analogies by cuing within-domain products, in order to identify problems with these previous exemplars that newly generated solutions also possess. Such exploratory creative processes may lead to the identification of problems and issues that need to be addressed in the innovation process.

Spontaneous Analogical Transfer Is Rare for Distant Analogies, But Instructions to Make Connections Help

In analogical transfer, spontaneous access refers to accessing a source without receiving hints or instructions to use this source. Gick and Holyoak's (1980) classic study showed that, unless explicitly instructed to try to make a connection between the problems, subjects rarely discovered their similarity and used it to solve the radiation problem. This lack of an automatic transfer has been repeated many times since, leading some authors to argue that analogical access is not a spontaneous process (e.g., Anolli et al., 2001). However, as Ross, Ryan, and Tenpenny (1989) noted, studies have not generally shown that people never spontaneously access relevant information—only that in cases where they were expected to do so, they often do not. As noted in the previous section, superficial similarity between source and target (as in within-domain analogizing) is one way of ensuring spontaneous transfer. Therefore, spontaneous retrieval of between-domain sources is quite rare, although a few studies have attempted to document that it is not altogether absent (Ball, Ormerod, and Morley, 2004; Christensen and Schunn, 2005). One approach to increasing transfer between source and target in between-domain analogizing involves bypassing the spontaneous access part of analogies, and simply instructing or hinting to subjects that they should make connections between source and target (Gick et al., 1980; for a more recent study in design, see Casakin and Goldschmidt, 1999). This approach increases transfer between even highly dissimilar domains. There is also evidence that presenting analogies in spoken form increases the chance of retrieval over the written format, particularly at longer lags between cue and recall (Markman, Taylor, and Gentner, 2007).

The lack of experimental evidence for spontaneous access is surprising when viewed in the light of *in vivo* research. Dunbar (Blanchette and Dunbar, 2000, 2001; Dunbar, 1995, 1997, 2001a) has conducted several studies of analogical transfer in real-world science and politics using the *in vivo* methodology (see Dunbar, 1995, 1997), and he has found that scientists and politicians frequently access analogues spontaneously (e.g., Dunbar, 2001a)—even for analogues sharing mainly deeper structural features (see also Bearman, Ball, and Ormerod, 2002, for similar results on a task in management decision-making). Our own findings from engineering design show similar results (Christensen and Schunn, 2007). This research finding obviously stands in sharp contrast to the experimental findings; a contrast Dunbar called "the analogical paradox." For some reason, experimental research and real-world research seem to reach opposite conclusions concerning frequency of spontaneous access. Experiments on spontaneous access are supposed to be studying a simplified version of the real world, but several differences exist between the two research strains that could potentially explain the paradox. One potentially important contrast highlighted by Blanchette and Dunbar (2000, 2001; Dunbar, 2001a) is that in the real-world, experts generate their own analogies, while in the typical experimental laboratory, subjects are provided with specific analogue sources. However, regardless of whether experimental studies using random cues may find some evidence of spontaneous between-domain analogizing in experimental research, there is no reason to think that instructing subjects to access random cues will not still lead to higher transfer rates. These findings have consequences for tools of innovations in that if a cuing tool (for example, a random picture viewer) is showing between-domain sources meant to promote the use of problem-solving analogies, it cannot be expected that designers will make use of the sources to any great degree, unless explicitly instructed to do so. The access path between distant analogues simply has too much resistance to lead to spontaneous use. This is not to say that an uninstructed or "priming" version of a distant analogy–generator may never work (as some research does show that in some cases transfer does occur even here)—but simply to say that the transfer between distant analogues is greatly enhanced by asking designers to actively think of connections between target and source. Furthermore, if within-domain sources are used in combination with between-domain sources, it may be expected that within-domain analogues are retrieved far more, due to paths of less resistance in both the instructed and uninstructed cases.

Explorations Using Mental Simulation of End Users Lead to More Useful Products

Another frequently used creative process in design and science involves mentally simulating events and entities under changed circumstances in order to support reasoning, understanding, and prediction (Gentner, 2002). In more popular terms, mental simulations have also been referred to as "thought experiments," and one of the most famous examples is probably Einstein's anecdote of how imagining traveling trough space next to a beam of light helped him discover the special theory of relativity (Einstein, in Hadamard, 1945, p. 142). Although different and competing paradigms of mental models have been proposed (e.g., Forbus and Gentner, 1997; Gentner and Stevens, 1983; Johnson-Laird, 1983; Kahneman and Tversky, 1982; Kuipers, 1994), these theories are basically in agreement with a minimalist approach hypothesizing that, in certain problem-solving tasks, humans reason by constructing a mental model of the situations, events, and processes in working memory that in dynamic cases can be manipulated through simulation (Christensen and Schunn, 2009; Nersessian, 2002). An important feature of mental models is that they frequently permit mental simulation. A *mental simulation* refers to the sense of being able to dynamically "run" a simulation internally to observe the functioning and outcome of a system or device. "*Runability*" implies a sense of being able to simulate system behavior and predict outcomes even for situations where the subject has no previous experience. In innovation, the potential advantage of using mental model runs include being able to reason about how physical systems will operate under changed circumstances and with altered features, without having to resort to physically constructing such a system or device. Mental model runs allow quick and cheap ways of exploring and testing possible alternatives. This ability is particularly useful in creative domains where uncertainty is an inescapable part of the problem space, since the task involves constructing novelty. While a mental model is runable, the mental simulation is the actual "run" (Trickett and Trafton, 2002; 2007; Trickett, 2004; Trickett, Trafton, Saner, and Schunn, 2005), and these "run" instances are detectable in verbal protocols. The different paradigms of mental models basically agree that mental models are run under situations of uncertainty, in order to turn that uncertainty into approximate or imprecise answers or solutions.

We tested this basic assumption *in vivo* by examining the mental simulations of engineering designers (Christensen and Schunn, 2009). We found that mental simulations were very frequently employed in design, and that they primarily served the function of mentally exploring and testing ideas, concepts, and prototypes that had been generated. We found support for the basic assumptions that (1) simulations are run in situations of increased uncertainty, (2) the running of mental simulations in effect reduces uncertainty, and (3) the resulting representations following mental simulations have increased approximation. Furthermore, we found that there were at least two different kinds of mental simulation in engineering design: simulations of technical or functional aspects, and simulations of end-user usability, preference, and product interaction.

We found that the reference to prototypes had fewer technical or functional simulations compared with sketches or unsupported cognition, perhaps indicating the lower degree of technical uncertainty in prototypes. While both types of simulations (technical/functional and end-user) may reduce uncertainty, notably the latter has been linked to creative outcomes in the literature. Theories of user-centered design (e.g., Norman and Draper, 1986), user involvement in design (e.g., Kujala, 2003), usability (e.g., Rubin, 1994), and user-driven innovation (e.g., von Hippel, 2005) all agree that considering or involving the end-user in design processes has the possibility of improving the usefulness of the resulting product. Usefulness is one of the defining characteristics of creative products (Mayer, 1999). In examining the impact on the resulting design of imagining end-users, Dahl, Chattopadhyay, and Gorn (1999) found that instructing designers to include the customer in the imagination's visual imagery during the design process has a greater positive effect on the usefulness of the designs produced than including the customer in memory's visual imagery, as evaluated by the target customers themselves. These findings point towards the theory that cuing random end-user information during innovative processes may lead designers to increasingly mentally simulate end-user preferences, usability, and product interactions, in order to explore and test the usefulness of the current pre-inventive structure. Insofar as these simulations point towards problems and potentials in the product, such increased end-user simulations may in effect lead to more useful products as evaluated by both designers and the end-users themselves.

A Model for Blind Cues, Cognitive Processes and Functions, and Creative Outcome

The findings from studies of analogical transfer and mental simulation led us to develop the following model of the relationship between random environmental cue categories, the cognitive processes of analogy and simulations, and the hypothesized creative outcome in terms of expected changes in originality or usefulness of the resulting product (see Fig. 3–2). Currently the model consists of three categories of random cues: within-domain products, between-domain products, and end-users. As more creative cognitive processes and their functions are examined, we expect that more categories will be specified, and that further restrictions for the current categories will be put forth. First, random between-domain cues will lead to increased between-domain analogizing primarily if subjects are instructed to make connections. Insofar as the analogies serve problem-solving purposes, this should lead to increased product originality. Second, random within-domain cues will lead to increased within-domain analogizing even when cues are presented without instructions. But higher levels of within-domain analogizing are expected with instruction to make connections. Due to property transfer, these close analogies will have a negative impact on the originality of the outcome in problem-solving instances, although they may also increase usefulness at the same time. In problem-identifying or exploratory instances, these close analogies should lead only to increased levels of outcome usefulness. Third, random end-user cues will lead to greater amounts of end-user simulations. The exploratory nature of end-user mental simulations will lead to considering usability and user preferences more, and thus to higher levels of product usefulness (as evaluated by the end-user).

A recent test of aspects of this model related to outcome usefulness (Christensen, under review) asked design students to design a new product within medical plastics, while being exposed to random images from these

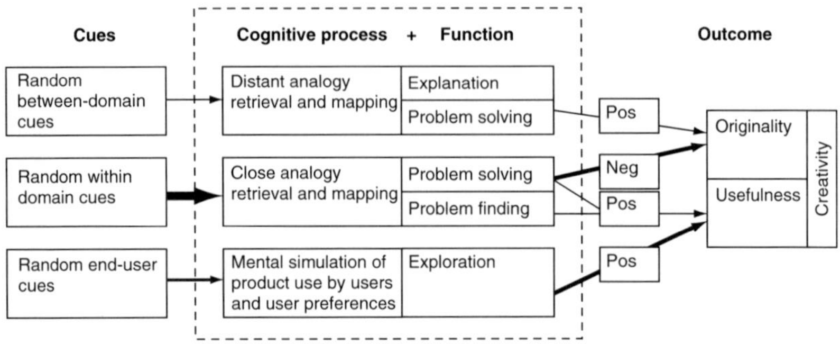

Figure 3–2 A model for the relation between random cue categories, creative cognitive processes and functions, and creative outcomes.

cue categories. We collected about 1,000 random pictures from both photo-databases with general content, and photo-databases with pictures from medical plastics, and coded each picture for category (within-domain; between-domain; end-users; other people; a control group viewed abstract art). Design students were then given 30 minutes to solve the design task, while being exposed to 60 random pictures from these cue categories. Following the experiment, end-users were asked to rate their willingness-to-use the solutions while blind to conditions. We also measured the design solutions for the amount of within-domain property transfer. In support of the model, we found that pictures of end-users did lead to improved ratings of willingness to use the resulting product. Furthermore, cues of within-domain products lead to increased transfer of within-domain properties, leading also to increased evaluations of willingness to use the final design. The experiment thus illustrated two different paths to increased outcome usefulness. Furthermore, support was found for the hypothesis that the within-domain cues were overshadowing the effect of end-user cues, when both categories were employed.

Practical Implications

Random environmental cues can be used to support creative processes, but the particular processes and their functions need to be considered before it can be hypothesized what the resulting impact will be on creative outcomes.

It is possible for the practitioner to pre-select random cues in certain categories and expect them to lead to improved creativity. However, in some cases they may also harm creativity—especially the originality of the outcome, if proper cue categories are not selected for support of the right creative processes. Some of the categories to be used in innovation can be seen in Table 3–1.

Table 3–1 Recommendations for Cue Categories.

Cue category	Recommendation
Within-domain products	This category will tend to lead designers to think in terms of local analogical solutions, and generate products that share a fair amount of elements with the past/cued solutions. In some cases this may affect the originality of the resulting product negatively. The effect of using this category seems to be rather potent, and it tends to overshadow some of the other cuing categories below. It is thus recommended that this category be used in isolation, without cross-cuing with the other categories. It is tentatively suitable for coming up with (generating) less-than-original solutions to problems, or for supporting exploratory processes of already generated solutions by means of identifying problems with these solutions based on previous knowledge. Because of the potency, the cuing category may be used either actively (as in instructing participants to try to relate to the cues) or passively (as in presenting cues without instructions during regular innovation processes).
Between-domain products	This category may in some cases lead to more original products by means of between domain analogizing. Due to low levels of shared superficial similarity with the problem at hand, the cues will seem less interesting and less related to the designers, unless explicitly instructed to make the connection. As such, the category should be used actively (i.e., instructed), although some research has shown that a small effect is also possible without instruction. Furthermore, if within-domain cues are present, this effect will be diminished due to less resistance in accessing those cues. Cuing between-domain products should primarily be used for generative problem-solving purposes, requiring greater originality in the solution. Further, distant analogies serve a natural explanatory purpose.

(Continued)

Table 3–1 (Continued)

Cue category	Recommendation
End-users	This category is notably useful in product design where there is an identifiable end-user. Cues of end-users may lead to greater amounts of mental simulation of end-users interacting with the innovation, potentially creating more useful products, as evaluated by design experts and the users themselves. It can be used either actively or passively, but it should not be used in combination with the overshadowing "within-domain product" category. Serves primarily exploratory functions of already generated designs. The use of this category does not limit the originality of the resulting design. The solution can be expected to be more useful and creative.

Conclusion

Cues are inherently neither good nor bad. But contextual information can be strategically selected and presented during creative processes to enhance the probability of ending up with an original and useful product. The problem is this: We want a random flux of information to inspire us and lead us along unexpected and potentially fruitful paths in creativity. But we do not want this flux to be misleading us along unfruitful paths. So a restricted randomization seems in order. However, in blind creative processes we do not know beforehand where we will end up. We have no *a priori* insight into which are the fruitful paths and which are the unfruitful ones until we have actually walked along them. The proposal here is that it is possible, if we examine how creative cognition works, to restrict the pool of random stimuli to increase the opportunity for great novelty, and decrease the probability of misleading failures. We have looked at the potential impact of three broad categories of cues, their impact on creative cognitive processes, and the expected outcomes. By setting limits to randomness, it is possible to exclude cues that would have promoted processes that may harm originality (such as property-transfer effects and other reproductive thinking processes), while at the same time to enhance processes that may lead to original and useful products.

Restricted randomness is not the same as algorithmic searches or the "working-out" of the typically well-defined problem, since, as noted by Cziko (1998, p. 207), "any adaptive constraint put on the current generation of blind variations does not make the resulting variations 'sighted,' 'smart,' or any less blind." Rather, by analogy with the leather shields placed over horses' eyes in order to restrict their vision, we are "putting blinkers on a blind man," by asking him to generate new variations from the road ahead, rather than from the road behind. The blind man still needs to walk the walk—but hopefully it will carry him into original and uncharted territory. Once there, the blind man needs to use his yardstick to scrutinize, test, and select the best possible ideas he has generated. Categories of environmental cues may help provide both the blinkers and the yardstick, by strategically supporting creative cognitive processes.

Creative search is still done without knowing where precisely you will end up in the infinite land of new ideas, but wherever the process may take you, the blinkers will ensure that it will probably not be in the land of the well-knowns. And the yardstick may ensure that the weird variations are quickly left behind, while the winning variations are selectively adopted.

References

Anolli, L., Antonietti, A., Crisafulli, L., and Cantoia, M. (2001). Accessing source information in analogical problem solving. *Quarterly Journal of Experimental Psychology: Human Experimental Psychology, 54A*, 237–261.

Ball, L. J., Ormerod, T. C., and Morley, N. J. (2004). Spontaneous analogising in engineering design: A comparative analysis of experts and novices. *Design Studies, 25*, 495–508.

Bearman, C. R., Ball, L. J., and Ormerod, T. C. (2002). An exploration of real-world analogical problem solving in novices. In W. Gray and C. D. Schunn (Eds.), *Proceedings of the 24th annual conference of the Cognitive Science Society* (pp. 101–106). Fairfax, Va.: Cognitive Science Society.

Beveridge, M., and Parkins, E. (1987). Visual representation in analogical problem solving. *Memory and Cognition, 15*, 230–237.

Blanchette, I., and Dunbar, K. (2000). How analogies are generated: The roles of structural and superficial similarity. *Memory and Cognition, 28*, 108–124.

Blanchette, I., and Dunbar, K. (2001). Analogy use in naturalistic settings: The influence of audience, emotion, and goals. *Memory and Cognition, 29*, 730–735.

Bonnardel, N., and Marmèche, E. (2004). Evocation processes by novice and expert designers: Towards stimulating analogical thinking. *Creativity and Innovation Management, 13*, 176–186.

Brown, A. S., and Murphy, D. R. (1989). Cryptomnesia: Delineating inadvertent plagiarism. *Journal of Experimental Psychology: Learning, Memory, and Cognition, 15,* 432–442.

Campbell, D. T. (1960). Blind variation and selective retentions in creative thought as in other knowledge processes. *Psychological Review, 67,* 380–400.

Casakin, H. (2004). Visual analogy as a cognitive strategy in the design process: Expert versus novice performance. *Journal of Design Research, 4.*

Casakin, H., and Goldschmidt, G. (1999). Expertise and the use of visual analogy: implications for design education. *Design Studies, 20,* 153–175.

Christensen, B. T. (under review). Images of users and products shown during new product design increase users' willingness-to-use the innovation.

Christensen, B. T., and Schunn, C. D. (2005). Spontaneous access and analogical incubation effects. *Creativity Research Journal, 17,* 207–220.

Christensen, B. T., and Schunn, C. D. (2007). The relationship between analogical distance to analogical function and pre-inventive structure: The case of engineering design. *Memory and Cognition 35 (1),* 29–38.

Christensen, B. T., and Schunn, C. D. (2009). The role and impact of mental simulation in design. *Applied Cognitive Psychology, 23,* 327–344.

Christiaans, H., and Andel, J. V. (1993). The effects of examples on the use of knowledge in a student design activity: The case of the "flying Dutchman." *Design Studies, 14,* 58–74.

Cziko, G. A. (1998). From blind to creative: In defense of Donald Campbell's selectionist theory of human creativity. *Journal of Creative Behavior, 32,* 192–209.

Dahl, D. W., Chattopadhyay, A., and Gorn, G. J. (1999). The use of visual mental imagery in new product design. *Journal of Marketing Research, 36,* 18–28.

Dahl, D. W., and Moreau, P. (2002). The influence and value of analogical thinking during new product ideation. *Journal of Marketing Research, 39,* 47–60.

De Bono, E. (1975). *The uses of lateral thinking.* New York: Harper and Row.

De Bono, E. (1992). *Serious creativity: Using the power of lateral thinking to create new ideas.* Toronto: HarperCollins Publishers Ltd.

Dunbar, K. (1995). How scientists really reason: Scientific reasoning in real-world laboratories. In R. J. Sternberg and J. E. Davidson (Eds.), *The nature of insight* (pp. 365–395). Cambridge, Mass.: The MIT Press.

Dunbar, K. (1997). How scientists think: On-line creativity and conceptual change in science. In T. B. Ward, S. M. Smith, and J. Vaid (Eds.), *Creative thought: An investigation of conceptual structures and processes* (pp. 461–493). Washington, D.C.: American Psychological Association.

Dunbar, K. (2001a). The analogical paradox: Why analogy is so easy in naturalistic settings yet so difficult in the psychological laboratory. In D. Gentner, K. J. Holyoak, and B. N. Kokinov (Eds.), *The analogical mind: Perspectives from cognitive science* (pp. 313–334). Cambridge, Mass.: The MIT Press.

Dunbar, K. (2001b). What scientific thinking reveals about the nature of cognition. In K. Crowley, C. D. Schunn, and T. Okada (Eds.), *Designing for science: Implications from everyday, classroom, and professional settings* (pp. 115–140). Mahwah, N.J.: Lawrence Erlbaum Associates.

DUNBAR, K., and BLANCHETTE, I. (2001). The in vivo/in vitro approach to cognition: The case of analogy. *Trends in Cognitive Sciences, 5,* 334–339.

DUNCKER, K. (1945). *On problem solving.* Westport, Conn.: Greenwood Press.

FINKE, R. A., WARD, T. B., and SMITH, S. M. (1992). *Creative cognition: Theory, research, and applications.* Cambridge, Mass.: MIT Press.

FIRESTIEN, R. L. (1996). *Leading on the creative edge. Gaining competitive advantage through the power of creative problem solving.* Colorado Springs, Colo.: Piñon Press.

FORBUS, K. D., and GENTNER, D. (1997). Qualitative mental models: Simulations or memories? In *Proceedings of the Eleventh International Workshop on Qualitative Reasoning,* Cortona, Italy.

FORBUS, K. D., GENTNER, D., and LAW, K. (1994). Mass. C/FAC: A model of similarity-based retrieval. *Cognitive Science, 19,* 141–205.

GENTNER, D. (1998). Analogy. In W. Bechtel and G. Graham (Eds.), *A companion to cognitive science* (pp. 107–113). Malden, Mass.: Blackwell Publishing.

GENTNER, D. (2002). Psychology of mental models. In N. J. Smelser and P. B. Bates (Eds.), *International encyclopedia of the social and behavioral sciences* (pp. 9683–9687). Amsterdam: Elsevier.

GENTNER, D., RATTERMANN, M. J., and FORBUS, K. D. (1993). The roles of similarity in transfer: Separating retrievability from inferential soundness. *Cognitive Psychology, 25,* 524–575.

GENTNER, D., and STEVENS, A. (1983). *Mental models.* Hillsdale, N.J.: Lawrence Erlbaum.

GHISELIN, B. (1954). *The creative process: A symposium.* Berkeley: University of California Press.

GICK, M. L., and HOLYOAK, K. J. (1980). Analogical problem solving. *Cognitive Psychology, 12,* 306–355.

GICK, M. L., and HOLYOAK, K. J. (1983). Schema induction and analogical transfer. *Cognitive Psychology, 15,* 1–38.

GOLDSCHMIDT, G. (2001). Visual analogy: A strategy for design reasoning and learning. In C. M. Eastman, W. M. McCracken, and W. C. Newstetter (Eds.), *Design knowing and learning: Cognition in design education* (pp. 199–220). Amsterdam: Elsevier.

HOLYOAK, K. J., and KOH, K. (1987). Surface and structural similarity in analogical transfer. *Memory and Cognition, 15,* 332–340.

HOLYOAK, K. J., and THAGARD, P. (1995). *Mental leaps: Analogy in creative thought.* Cambridge, Mass.: MIT Press.

JAARSVELD, S., and VAN LEEUWEN, C. (2005). Sketches from a design process: Creative cognition inferred from intermediate products. *Cognitive Science, 29,* 79–101.

JANSSON, D. G., and SMITH, S. M. (1991). Design fixation. *Design Studies, 12,* 3–11.

JOHNSON-LAIRD, P. N. (1983). *Mental models.* Cambridge, U.K.: Cambridge University Press.

JOHNSON-LAIRD, P. N. (1989). Analogy and the exercise of creativity. In S. Vosniadou and A. Ortony (Eds.), *Similarity and analogical reasoning* (pp. 313–331). New York: Cambridge University Press.

KAHNEMAN, D., and TVERSKY, A. (1982). The simulation heuristic. In D. Kahneman, P. Slovic, and A. Tversky (Eds.), *Judgement under uncertainty. Heuristics and biases.* Cambridge, U.K.: Cambridge University Press.

KUIPERS, B. (1994). *Qualitative reasoning.* Cambridge, Mass.: The MIT Press.

KUJALA, S. (2003). User involvement: A review of the benefits and challenges. *Behaviour and Information Technology, 22,* 1–16.

LUCHINS, A. S. (1942). Mechanisation in problem solving. The effect of *Einstellung. Psychological Monographs, 54,* no. 248.

MACCRIMMON, K. R., and WAGNER, C. (1994). Stimulating ideas through creativity software. *Management Science, 40,* 1514–1532.

MAIER, N. R. F. (1931). Reasoning in humans, II: The solution of a problem and its appearance in consciousness. *The Journal of Comparative Psychology, 8,* 181–194.

MARKMAN, A. B., TAYLOR, E, and GENTNER, D. (2007). Auditory presentation leads to better analogical retrieval than written presentation. *Psychonomic Bulletin and Review, 14(6),* 1101–1106.

MARSH, R. L., and BOWER, G. H. (1993). Eliciting cryptomnesia: Unconscious plagiarism in a puzzle task. *Journal of Experimental Psychology: Learning, Memory, and Cognition, 19,* 673–688.

MARSH, R. L., and LANDAU, J. D. (1995). Item availability in cryptomnesia: Assessing its role in two paradigms of unconscious plagiarism. *Journal of Experimental Psychology: Learning, Memory, and Cognition, 21,* 1568–1582.

MARSH, R. L., LANDAU, J. D., and HICKS, J. L. (1996). How examples may (and may not) constrain creativity. *Memory and Cognition, 24,* 669–680.

MARSH, R. L., LANDAU, J. D., and HICKS, J. L. (1997). Contributions of inadequate source monitoring to unconscious plagiarism during idea generation. *Journal of Experimental Psychology: Learning, Memory, and Cognition, 23,* 886–897.

MARSH, R. L., WARD, T. B., and LANDAU, J. D. (1999). The inadvertent use of prior knowledge in a generative cognitive task. *Memory and Cognition, 27,* 94–105.

MAYER, R. E. (1999). Fifty years of creativity research. In R. J. Sternberg (Ed.), *Handbook of creativity* (pp. 449–460). Cambridge, U.K.: Cambridge University Press.

MEDNICK, S. A. (1962). The associative basis of the creative process. *Psychological Review, 69,* 220–232.

METCALFE, J., and WIEBE, D. (1987). Intuition in insight and non-insight problem solving. *Memory and Cognition, 15,* 238–246.

NERSESSIAN, N. J. (2002). The cognitive basis of model-based reasoning in science. In P. Carruthers and S. Stich (Eds.), *Cognitive basis of science* (pp. 133–153). New York: Cambridge University Press.

NEWELL, A., and SIMON, H. A. (1972). *Human problem solving.* Englewood Cliffs, N.J.: Prentice-Hall.

NORMAN, D. A., and DRAPER, S. W. (1986). *User Centered System Design; New Perspectives on Human–Computer Interaction.* Mahwah, N.J.: Lawrence Erlbaum Associates.

NOVICK, L. R. (1988). Analogical transfer, problem similarity, and expertise. *Journal of Experimental Psychology: Learning, Memory, and Cognition, 14,* 510–520.

OSBORN, A. F. (1963). *Applied imagination* (3rd rev. ed.). New York: Charles Scribner's Sons.

PERKINS, D. N. (2000). *Archimedes' bathtub: The art and logic of breakthrough thinking.* New York: W.W. Norton and Company.

PURCELL, A. T., and GERO, J. S. (1992). Effects of examples on the results of a design activity. *Knowledge-Based Systems, 5,* 82–91.

Roozenburg, N. F. M., and Eekels, J. (1996). *Product design: Fundamentals and methods.* Chichester, U.K.: John Wiley and Sons.

Ross, B. H. (1987). This is like that: The use of earlier problems and the separation of similarity effects. *Journal of Experimental Psychology: Learning, Memory, and Cognition, 13,* 629–639.

Ross, B. H. (1989). Distinguishing types of superficial similarities: Different effects on the access and use of earlier problems. *Journal of Experimental Psychology: Learning, Memory, and Cognition, 15,* 456–468.

Ross, B. H., Ryan, W. J., and Tenpenny, P. L. (1989). The access of relevant information for solving problems. *Memory and Cognition, 17,* 639–651.

Rubin, J. (1994). *Handbook of usability testing: How to plan, design, and conduct effective tests.* New York: John Wiley and Sons.

Runco, M. A. (1985). Reliability and convergent validity of ideational flexibility as a function of academic achievement. *Perceptual and Motor Skills, 61,* 1075–1081.

Shepard, R. N. (1978). Externalization of mental images and the act of creation. In B. S. Randawa and W. E. Cofman (Eds.), *Visual learning, thinking, and communication* (pp. 133–189). New York: Academic Press.

Simon, H. A., and Hayes, J. R. (1976). The understanding process: Problem isomorphs. *Cognitive Psychology, 8,* 165–190.

Simonton, D. K. (2003). Scientific creativity as constrained stochastic behavior: The integration of product, person, and process perspectives. *Psychological Bulletin, 129,* 475–494.

Smith, S. M. (1995a). Fixation, incubation, and insight in memory and creative thinking. In S. M. Smith, T. B. Ward, and R. A. Finke (Eds.), *The creative cognition approach* (pp. 135–156). Cambridge, Mass.: The MIT Press.

Smith, S. M. (1995b). Getting into and out of mental ruts: A theory of fixation, incubation, and insight. In R. J. Sternberg and J. E. Davidson (Eds.), *The nature of insight* (pp. 229–251). Cambridge, Mass.: The MIT Press.

Smith, S. M., Ward, T. B., and Schumacher, J. S. (1993). Constraining effects of examples in a creative generations task. *Memory and Cognition, 21,* 837–845.

Terninko, J., Zusman, A., and Zlotin, B. (1998). *Systematic innovation: An introduction to TRIZ.* Boca Raton, Fla.: St. Lucie Press.

Trickett, S. B. (2004). Movies-in-the-mind: The instantiation and use of conceptual simulations in scientific reasoning. Unpublished doctoral dissertation from George Mason University.

Trickett, S. B., and Trafton, J. G. (2002). The instantiation and use of conceptual simulations in evaluating hypotheses: Movies-in-the-mind in scientific reasoning. In *Proceedings of the 24th Annual Conference of the Cognitive Science Society* (pp. 878–883). Mahwah, N.J.: Erlbaum.

Trickett, S. B., Trafton, J. G., Saner, L., and Schunn, C. D. (2005). "I don't know what's going on there": The use of spatial transformations to deal with and resolve uncertainty in complex visualizations. In M. C. Lovett and P. Shah (Eds.), *Thinking with data.* Mahwah, N.J.: Erlbaum.

Trickett, S. B., and Trafton, J. G. (2007). "What if. . ." : The use of conceptual simulations in scientific reasoning. *Cognitive Science, 31,* 843–875.

Von Hippel, E. (2005). *Democratizing innovation.* Cambridge: The MIT Press.

Vosniadou, S., and Ortony, A. (1989). Similarity and analogical reasoning: A synthesis. In S. Vosniadou and A. Ortony (Eds.), *Similarity and analogical reasoning* (pp. 1–7). New York: Cambridge University Press.

Ward, T. B. (1994). Structured imagination: The role of category structure in exemplar generation. *Cognitive Psychology, 27,* 1–40.

Ward, T. B. (1995). What's old about new ideas? In S. M. Smith, T. B. Ward, and R. A. Finke (Eds.), *The creative cognition approach* (pp. 157–178). Cambridge, Mass.: The MIT Press.

Ward, T. B. (1998). Analogical distance and purpose in creative thought: Mental leaps versus mental hops. In K. J. Holyoak, D. Gentner, and B. N. Kokinov (Eds.), *Advances in analogy research: Integration of theory and data from the cognitive, computational, and neural sciences.* Sofia, Bulgaria: New Bulgarian University.

Ward, T. B., Patterson, M. J., Sifonis, C. M., Dodds, R. A., and Saunders, K. N. (2002). The role of graded category structure in imaginative thought. *Memory and Cognition, 30,* 199–216.

Ward, T. B., Smith, S. M., and Vaid, J. (1997). Conceptual structures and processes in creative thought. In T. B. Ward, S. M. Smith, and J. Vaid (Eds.), *Creative thought: An investigation of conceptual structures and processes* (pp. 1–27). Washington, D.C.: American Psychological Association.

Wertheimer, M. (1959). *Productive thinking* (enlarged ed.). New York: Harper and Row, Publishers.

CHAPTER 4

THINKING WITH SKETCHES

BARBARA TVERSKY

MASAKI SUWA

Why Sketch?

Designers sketch. One reason they sketch is that they design things that can be seen. A sketch can resemble what the designer wants to create. Unlike the contents of the imagination, a sketch can be seen. Thus, sketches serve to amplify a designer's imagination and relieve limited-capacity working memory. Sketches map on paper things that exist in the world or the imagination and the relations among them, spatial or abstract, to elements and relations on paper: a natural mapping. They can be used to convey concepts that are literally spatial, such as objects, buildings, and environments, as well as concepts that are metaphorically spatial, such as information systems, organization charts, and family trees.

Models can convey objects and spaces as well, perhaps more so than sketches, since models are three-dimensional. Both have a place in design. Early in the design process, sketches have advantages over models, especially when the designer is considering many alternatives, which may be vague or partial. Sketches are just that, *sketchy*; for example, they can represent incomplete objects as blobs, or incomplete connections as wavy lines, so that a designer can consider general configurations before committing to particular connections and specific shapes. Models demand completeness.

Sketching is faster than model-building, and kinder to trial-and-error and revision. Sketches are easier to create and easier to revise. Sketches are two-dimensional, and thinking in two dimensions is easier than thinking in three (e.g., Gobert, 1999; Shah and Carpenter, 1995). Sketches readily enable expressing separate parts, different perspectives, and various scales so the designer can focus on each part, perspective, or scale, viewing them separately or together. A designer can use a sketch to focus on certain aspects of the design, ignoring others that may distract from the problem at hand.

Because sketches are visible they can be inspected and reinspected, considered and reconsidered. Designers can discover new properties and relations from their sketches as they inspect them—properties and relations that emerge from the sketch but were not intentionally put there (e.g., Goel, 1995; Goldschmidt, 1994; Schon, 1983; Suwa, Tversky, Gero, and Purcell, 2001). But sketches can go beyond the visible. They can eliminate detail that is irrelevant and distracting in the service of capturing the essential. At the same time, sketches can exaggerate and even distort the essential. They can be enriched with words and other symbols, enhancing their meaning with ideas and properties not easily expressed pictorially. Yet designers of things that cannot be seen also sketch. Sketches can use elements and relations on paper to represent abstract elements and abstract relations. Designers of experiments and of assembly lines, both of which occur in time, sketch possible sequences of events. Extending sketches from space to time is a natural step, as temporal events are described in part using the language of space, for example, *before* and *after, forward* and *back.* But designers of abstract ideas such as corporate organizations and computer operating systems also sketch.

Why can't people work these things out in their heads? To some extent they, especially experts, can, but ideas, whether spatial or abstract, that are complex or detailed are likely to be too massive to hold in one's mind, especially if they need examination, manipulation, or revision—all processes crucial to design. The pragmatics of putting ideas on paper demands a degree of coherence, completeness, and consistency, serving as a test of design ideas. Finally, the public nature of sketches facilitates communicating ideas to others and collaborating with others (Heiser and Tversky, 2004; Heiser, Tversky, and Silverman, 2004). Sketches serve as an easy referent for words and gestures, so deictic expressions like *here* and *there* and *this part* and *that way* simultaneously make communication easier and more precise. In collaborations, they represent the ideas of the group, not of any individual, so all are committed to it. When we've barely begun to formulate a concept, sketches are useful because they externalize ideas, encourage coherence and completeness, allow expression of the vague as well as the specific, map large space to small, extract the crucial, enrich by annotation, make the abstract concrete, relieve limited working memory, facilitate information processing, encourage inference and discovery, and promote collaboration—and more (Tversky, 1999, 2001, 2005).

What Is the Nature of Sketches?

Design sketches are imprecise, at least at the beginning. They are tentative; they do not commit the designer to exact shapes or exact spatial relations. They use a limited vocabulary of abstract shapes whose meanings are suggested by their geometric or gestalt properties. In early design in architecture, for example, blobs can stand for structures, buildings or rooms, and lines for the paths or corridors that connect them. Blobs are used to represent concepts we think of as three-dimensional (turned two-dimensional on paper) and lines to represent concepts we think of as two-dimensional (turned one-dimensional on paper).

Diagrams also use a limited vocabulary of shapes, but they contrast with sketches in being exact and definitive. In sketches, the tentative nature of shapes and spatial relations is directly suggested by irregularities, by imperfections, by inexact tracings and retracings. By contrast, in diagrams such as circuit and molecular diagrams, for instance, shapes and lines tend to be symmetric or regular or straight. Whereas sketches are often meant to be suggestive, tentative, ambiguous, and open to reinterpretation, diagrams, especially explanatory ones, are meant to be clear and unambiguous, in order to avoid ambiguities and misinterpretations.

An example of a type of diagram that has become conventionalized through use is the route map. Route maps are meant for clear, unambiguous communication, not for creative design. Although route maps could be analog, they are not. In fact, they seem to schematize environments exactly the way human memory does, by straightening roads, making turns into right angles and roads parallel, by distorting distances (e.g., Tversky, 1981, 2005). An analysis of a corpus of route maps students spontaneously produced to guide a traveller revealed a small number of elements with quite specific meanings. These elements can be concatenated in specific ways to convey a multitude of routes. That is, route maps have a semantics and a syntax (Tversky and Lee, 1998, 1999). The semantics consisted of what might be called *graphemes*: for turns, L's, T's, and +'s; for straight paths, lines; for curved paths, arcs of circles; for landmarks, street names or blobs.

Significantly, the semantics for verbal route descriptions revealed parallel elements, "make a," "take a," or "turn" for turns; "go down" for straight paths; "follow around" for curved paths; names for landmarks. Each set of semantic elements, descriptive or depictive, forms an essentially complete semantics of routes. In one study, groups of participants were assigned either the visual or the verbal vocabulary and asked to use the vocabulary to construct a wide range of routes, short or long, simple or complex (Tversky and Lee, 1998). Participants were told they could add elements if needed, but very few did; in other words, the sets of elements were virtually sufficient. The semantics of gestures used in describing routes include the same elements (Tversky, Heiser, Lee, and Daniel, 2009). The parallels between the semantics and syntax of depictions, descriptions, and gestures of routes suggest that they derive from same underlying mental model.

At an even more abstract level, the primary elements in route maps indicate concepts that are thought of as points, as lines, as areas, and as volumes. Design sketches also use these elements. Interestingly, a similar tripartite distinction is one that Talmy (1984) has proposed to characterize the language of space and time: one-dimensional points, two-dimensional areas, and three-dimensional volumes. We say the group will meet "at the corner at 1:45," point-like spatial and temporal concepts. The hike will go "from the Capitol to Barton Pond from 2 to 4," both like lines or areas. It will take place "in two days' time in Austin," both volume-like concepts. These conceptual distinctions have metaphorical extensions: someone can be *at* a crisis but *on* top of things, so not *in* a panic.

The comparison of the semantics and syntax used to convey routes in descriptions, depictions, and gestures, then, have revealed the underlying mental models people use for routes. A route consists of landmarks and paths, nodes and edges, turns and progressions. Exact distances and directions are not important, as they can be inferred from context, and neither are the regions not along the route. The underlying mental model and the graphic devices used to convey it can serve as cognitive design principles for creating sketches or diagrams that are useful and effective. This program of eliciting mental models from depictions and descriptions, extracting from cognitive design principles from them, and incorporating the cognitive design principles into algorithms to generate diagrams on demand has been successfully applied to both route maps and assembly instructions (Tversky et al., 2007). The program can be adopted for other domains. The productions of depictions and descriptions (and also gestures; see Tversky, Heiser, Lee, and Daniel, 2009) simultaneously reveal the underlying mental models and suggest effective depictive and descriptive semantic elements and syntactic rules.

As design progresses, vocabularies and nuances grow and expand; sketches are typically enriched and articulated so that shapes of paths and regions, parts and wholes, are more specific, and can be recognized as such. Do and Gross (summarized in Do, 2005) have studied the visual vocabularies of architects as they expand. Even expanded and articulated, elements remain sketchy and schematic. For example, people may be represented as stick figures and rooms as rectangles with openings for doors.

How Are Design Sketches Used?

Early design sketches are even more inexact than route maps. One reason is that the designer hasn't yet committed to specifics. Another reason, intended or unintended, is that sketches, because they are ambiguous, support many interpretations. The ambiguity of design sketches, rather than promoting confusion, promotes innovation. Because they support many interpretations, early sketches can be used for discovery and reinterpretation to further the design. Schon (1983) has described this as a conversation designers have with their own sketches. The designer creates the sketch to represent one set of constraints, elements, and relations, but on re-examining the sketch, sees other elements, relations, and patterns (e.g. Goldschmidt, 1994; Suwa, Gero, and Purcell, 2000; Suwa and Tversky, 1997). These unintended discoveries advance the design. In one study, novice and experienced architects were asked to design a museum on a particular site (Suwa and Tversky, 1997). Their design sessions were filmed, and afterwards, the designers viewed their sessions and explained what they were thinking at each stroke of the pencil. Both novice and expert architects got new ideas from examining their own sketches. However, the expert architects were more likely to get functional ideas from their sketches. The novices discovered structural features and relations in their own sketches; arguably, these require little interpretation as the structural features and relations are "there" in the sketch, ready to be perceived. Experts, by contrast, could "see" functional features and relations in their sketches, for example, changes of light or flow of traffic. These functional features and relations are not directly visible in the sketch, but require complex inferences entailing expertise. Seeing function in structure in fact seems to be a hallmark of expertise: for example, in chess (Chase and Simon, 1973; de Groot, 1965) and in engineering diagrams (Heiser and Tversky, submitted). Expertise, then, promotes seeing function in form.

How Can Sketches Be Effectively Reinterpreted?

What leads to these reinterpretations, so crucial to advancing design? A detailed study of one expert architect revealed that most of his new ideas came when he perceived the elements of the sketch differently; that is, when he reconfigured them into a different pattern (Suwa, Gero, and Purcell, 2000). A new idea, in turn, allowed him to reconfigure the sketch yet again, so that a positive cycle ensued: perceptual reorganization generating new conceptions and new conceptions generating perceptual reorganizations.

Is this strategy of searching for new perceptual relations a general one? The next step was to turn from designers to undergraduates (Suwa, Tversky, Gero, and Purcell, 2001). We showed undergraduates a series of ambiguous but suggestive sketches, those in Figure 4–1. Their task, adapted from a procedure used by Howard-Jones (1998), was to generate as many new interpretations as they could think of for each, taking four minutes for each sketch in turn. Approximately two-thirds of them used a strategy of attending to the parts of each sketch, either focussing successively on different parts or attempting to rearrange the parts mentally, for the purpose of coming up with new interpretations. Those who adopted an attention-to-parts strategy were more successful than those who didn't. Those who attended to different parts came up with 45 interpretations, and those who rearranged parts produced 50 interpretations, both in contrast to the participants who did not perceptually reconfigure the sketch and who produced only an average of 27 interpretations.

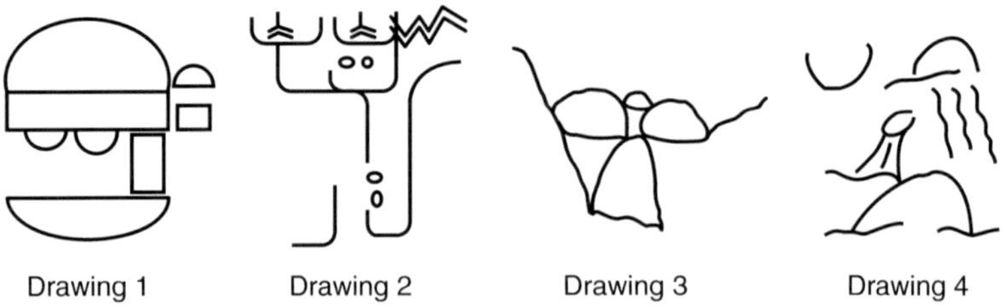

Figure 4–1 Four ambiguous drawings.

The parts-focus strategy was also effective against *fixation*, the plague of designers: getting stuck on a particular design and not being able to see alternatives. During the early phases of design, designers typically generate many ideas, but as they work and their designs become more constrained, they find it more and more difficult to see alternative solutions. The undergraduates who adopted the parts-focus strategy succeeded in producing more ideas in the second half of each session than those who did not.

Perhaps not surprisingly, practicing designers were more fluent at the task of generating new ideas and produced more of them than did the design students and laypeople (Suwa and Tversky, 2001, 2003). This suggests that experience promotes the required skills. The practicing designers reported a variety of ways to perceptually reconfigure the sketches, notably regrouping the parts and changing reference frames. In addition, they sometimes reversed figure and ground relations in the sketches. However, perceptual reorganization is only half of the process of coming up with new interpretations. Those interpretations must have meaning. To some extent, both the perceptual skill and the conceptual skill can be measured. The perceptual skill is measured by the embedded-figures test, in which participants' ability to see a simple geometric figure in a complex one is assessed. The conceptual skill is measured by an associative-fluency task in which participants' ability to find a meaningful association relating two unrelated words is assessed. The number of interpretations produced increased with each of these abilities independently. That is, those proficient in perceiving embedded figures and those high in associative fluency produced more interpretations, but the two abilities were not correlated.

Integrating these results suggests that actively reconfiguring sketches and finding meanings in them, termed *constructive perception*, promotes new design ideas and protects against fixation. The fact that designers are more proficient than laypersons suggests that the skill can be fostered. The fact that abstract ideas can be sketched suggests that constructive perception may have applications beyond the design of real objects and structures to the design of abstract objects and structures.

Implications

Design entails generating ideas and adapting them to users. This requires thinking broadly about possibilities and linking those possibilities to meaningful uses. This process is iterative, and facilitated by sketches. Sketches allow designers to express ideas both vague and developed, and then see their ideas, contemplate them, alter them, and refine them. This iterative process of constructing, examining, and reconstructing has been called a kind of "conversation" (Schon, 1983). Successful conversation with sketches depends on finding new perceptual configurations as well as new meanings, and connecting the configurations to the meanings, seeing function in form.

Designers report, and research supports, that sketches, even rudimentary and ambiguous ones, are helpful to design early on. These early sketches typically capture very general aspects of a design, using a limited range of domain-specific visual elements. As design progresses, sketches become more articulated. CAD/CAM tools are often avoided in early phases of design because they require or impose a completeness that is premature (e.g., Do, 2005; Hearst et al., 1996). There are ongoing efforts to adapt these findings to create tools that can facilitate both early and late processes of design (e.g., Do, 2005; Hearst et al., 1996). These tools try to facilitate design first by aiding sketching: recognizing the primitive elements, often completing and remembering them, and allowing them to be manipulated and replicated. The tools can also enhance design in ways that go beyond sketching, by retrieving examples that use similar elements or have similar goals so that the designer can use these as examples or analogies. These other examples can be related artifacts, such as other spiral staircases or buildings, or natural objects with similar shapes or goals, such as snails. By retrieving examples that are functionally similar as well as examples that are perceptually similar, these tools can aid both perceptual and functional aspects of constructive perception. A rich and relevant source of examples can increase innovation by providing the designer with ideas the designer might not otherwise consider. A broad range of new examples can also break fixation, a persistent problem for designers. These new tools have the potential not only to facilitate innovative design but also to make it more fun.

Acknowledgments

Gratitude to the following grants for partial support of some of the research: Office of Naval Research Grants NOOO14-PP-1-O649, N00014011071, and N000140210534, NSF Grant REC-0440103, NSF Grant IIS-0725223, the Stanford Regional Visual Analytics Center, and an Edinburgh-Stanford Link grant to Stanford University.

References

CHASE, W. G., and SIMON, H. A. (1973). The mind's eye in chess. In W. G. Chase (Ed.), *Visual information processing.* New York: Academic Press.

DE GROOT, A. D. (1965). *Thought and choice in chess.* The Hague: Mouton.

DO, E. Y.-L. (2005). Design sketches and sketch design tools. *Knowledge-Based Systems, 18*, 383–405.

GOBERT, J. D. (1999). Expertise in the comprehension of architectural plans. In J. Gero and B. Tversky (Eds.), *Visual and spatial reasoning in design* (pp. 185–205). Sydney, Australia: Key Centre of Design Computing and Cognition.

GOEL, V. (1995). *Sketches of thought.* Cambridge, Mass.: MIT Press.

GOLDSCHMIDT, G. (1994). On visual design thinking: The vis kids of architecture. *Design Studies, 15*(2), 158–174.

HEARST, M. A., GROSS, M. D., LANDAY, J. A., and STAHOVICH, T. F. (1996). Sketching intelligent systems. *IEEE Intelligent Systems, 13,* 10–19.

HEISER, J., and TVERSKY, B. (2004a). Characterizing diagrams produced by individuals and dyads. In T. Barkowsky (Ed.), *Spatial cognition: Reasoning, action, interaction* (pp. 214–223). Berlin: Springer-Verlag.

HEISER, J., TVERSKY, B., and SILVERMAN, M. (2004b). Sketches for and from collaboration. In J. S. Gero, B. Tversky, and T. Knight (Eds.), *Visual and spatial reasoning in design III* (pp. 69–78). Sydney, Australia: Key Centre for Design Research.

HEISER, J., and TVERSKY, B. (submitted). Mental models of complex systems: Structure and function.

HOWARD-JONES, P. A. (1998). The variation of ideational productivity over short time-scales and the influence of an instructional strategy to defocus attention. *Proceedings of Twentieth Annual Meeting of the Cognitive Science Society* (pp. 496–501). Hillsdale, N. J.: Lawrence Erlbaum Associates.

JANSSON, D. G., and SMITH, S. M. (1991). Design fixation. *Design Studies, 12*, 3–11.

SCHON, D. A. (1983). *The reflective practitioner.* New York: Harper Collins.

SHAH, P., and CARPENTER, P. A. (1995). Conceptual limitations in comprehending line graphs. *Journal of Experimental Psychology: General, 124*, 43–61.

SUWA, M., GERO, J., and PURCELL, T. (2000). Unexpected discoveries and S-invention of design requirements: Important vehicles for a design process. *Design Studies, 21*, 539–567.

SUWA, M., and TVERSKY, B. (1997). What do architects and students perceive in their design sketches? A protocol analysis. *Design Studies, 18*, 385–403.

SUWA, M., and TVERSKY, B. (2001). Constructive perception in design. In J. S. Gero and M. L. Maher (Eds.), *Computational and cognitive models of creative design V* (pp. 227–239). Sydney, Australia: University of Sydney.

SUWA, M., and TVERSKY, B. (2003). Constructive perception: A skill for coordinating perception and conception. In R. Alterman and D. Kirsh (Eds.), *Proceedings of the Cognitive Science Society Meetings.* Boston, MA: Cognitive Science Society.

SUWA, M., TVERSKY, B., GERO J., and PURCELL, T. (2001). Regrouping parts of an external representation as a source of insight. *Proceedings of the 3rd International Conference on Cognitive Science* (pp. 692–696). Beijing, China: Press of University of Science and Technology of China.

Talmy, L. (1984). How language structures space. In H. L. Pick, Jr., and L. P. Acredolo (Eds.), *Spatial orientation: Theory, research and application* (pp. 225–282). New York: Plenum.

Tversky, B. (1981). Distortions in memory for maps. *Cognitive Psychology, 13*, 407–433.

Tversky, B. (1999). What does drawing reveal about thinking? In J. S. Gero and B. Tversky (Eds.), *Visual and spatial reasoning in design* (pp. 93–101). Sydney, Australia: Key Centre of Design Computing and Cognition.

Tversky, B. (2001). Spatial schemas in depictions. In M. Gattis (Ed.), *Spatial schemas and abstract thought* (pp. 79–111). Cambridge, Mass.: MIT Press.

Tversky, B. (2005). Functional significance of visuospatial representations. In P. Shah and A. Miyake (Eds.), *Handbook of higher-level visuospatial thinking* (pp. 1–34). Cambridge, U.K.: Cambridge University Press.

Tversky, B., Agrawala, M., Heiser, J., Lee, P. U., Hanrahan, P., Phan, D., Stolte, C., Daniele, M.-P. (2007). Cognitive design principles for generating visualizations. In G. Allen (Ed.), *Applied spatial cognition: From research to cognitive technology*. Mahwah, N. J.: Erlbaum.

Tversky, B., Heiser, J., Lee, P., and Daniel, M.-P. (2009). Explanations in gesture, diagram, and word. In K. R. Coventry, T. Tenbrink, and J. Bateman (Eds.), *Spatial language and dialogue*. Oxford: Oxford University Press.

Tversky, B., and Lee, P. U. (1998). How space structures language. In C. Freksa, C. Habel, and K. F. Wender (Eds.), *Spatial cognition: An interdisciplinary approach to representation and processing of spatial knowledge* (pp. 157–175). Berlin: Springer-Verlag.

Tversky, B., and Lee, P. U. (1999). Pictorial and verbal tools for conveying routes. In C. Freksa and D. M. Mark (Eds.), *Spatial information theory: cognitive and computational foundations of geographic information science* (pp. 51, 64). Berlin: Springer.

CHAPTER 5

SUPPORTING INNOVATION BY PROMOTING ANALOGICAL REASONING

ARTHUR B. MARKMAN

KRISTIN L. WOOD

JULIE S. LINSEY

JEREMY T. MURPHY

JEFFREY P. LAUX

HUMAN behavior contains a striking mix of habit and creative behavior. On one hand, much of what we do in life is routine. We tend to take the same route to work each day. We sit in the same seats in classes and meetings. We purchase the same products at the supermarket. On the other hand, our daily life is marked by language use in which we produce novel sentences in new contexts, communicating our thoughts with sentences we have never uttered before. While many of our behaviors are routine, we are also capable of adapting to new circumstances flexibly.

Much of our everyday behavior—both the habitual and the productive—feels effortless. In contrast, innovation settings are often effortful and frustrating. Consequently, we are prone to think that innovation requires cognitive processes quite different from those involved in our daily behavior.

We suggest that the feeling of frustration and effort involved in innovation settings arises from an inability to retrieve relevant knowledge that suggests a solution to the current problem. That is, a critical bottleneck in innovative problem solving is the ability of a problem solver (or problem-solving team) to identify prior instances or principles that facilitate problem solving. Once we understand people's strengths and weaknesses in their ability to retrieve background knowledge, we can develop tools that improve these abilities.

In order to develop these ideas further, we begin with a discussion of the role of background knowledge in creative problem solving. This discussion will focus on the way problems are categorized and the importance of analogy in problem solving. This presentation naturally leads to a focus on limitations of people's ability to retrieve analogies. We discuss research that examines methods to improve people's ability to retrieve analogies. Finally, we examine important avenues for tools that are capable of extending people's ability to use prior knowledge to solve new problems.

Analogy and Problem Solving

Cognitive science typically takes one of two approaches to studying problem solving. One is a problem space approach, and the second is a background knowledge approach. In the problem space view, problems arise when people have a goal that they must achieve, and they have a set of steps or operations that are available to solve the problem, but the sequence of steps or the set of relevant operations is not known (e.g., Newell, 1990; Newell and Simon, 1963). In the background knowledge view, a central method for solving a new problem is to find a prior problem that bears important similarities to the current problem and then to adapt the solution to the old problem to the new situation (e.g., Polya, 1945). The background knowledge might be an analogous situation, or it might be a known specific case from a similar domain (Gick and Holyoak, 1980; Kolodner, 1993).

It is easy to see the role of background knowledge in innovations retrospectively. For example, barbed wire was modeled on briar bushes that were grown in the west to provide livestock barriers (Basalla, 1988). De Mestral is known to have invented Velcro® after seeing burrs sticking to the fur of his dog. Weisberg (this volume) discusses the evolution of ideas over time in the works of creative individuals. Christensen and Schunn (2007) document the uses of analogy by innovators working in the domain of medical plastics.

The critical issue for promoting innovation, however, is to understand the way people come to recognize that knowledge they have in one domain is going to be useful to solve the current problem. That is, how can we use analogy prospectively? To address this question, we first have to give a brief summary of what is known about similarity and analogy. This theoretical basis will ground our discussion about the difficulties of using analogies and the bases for tools to support innovation.

Analogical Reasoning

Much psychological research has examined people's ability to form and use analogies. This work has established a set of basic principles for the psychology of analogy that are generally agreed upon by researchers (Gentner, 1983; Gentner, Holyoak, and Kokinov, 2001; Holyoak and Thagard, 1995). This work suggests that analogies involve finding parallel sets of relationships between two domains. We can illustrate this idea with the analogy between an inflatable mattress and water weights described in Figure 5–1.

Inflatable Mattress:	Water Weights:
Goal: Mattress that can be easily packed *Obstacle*: Regular mattress is heavy *Solution*: Replace mattress filling with air Air-filled mattress supports body Mattress can be filled on site	*Goal*: Weights that can be easily packed *Obstacle*: Weight sets are heavy *Solution*: Replace weights with water-filled bag Water is heavy Weights can be filled on site

Figure 5–1 Analogy between an inflatable mattress and water-filled weights.

An inflatable mattress is used by campers to provide a comfortable surface for sleeping. The mattress is a plastic shell that is inflated with air when it is set up. Water weights are a workout set consisting of inflatable plastic pouches connected to bars. The plastic pouches are empty when packed, but can be filled with water to allow travelers to lift weights on a trip.

These devices are not particularly similar either in the way they look, their specific functions, or the way they operate. An inflatable mattress is large; water weights are small. Mattresses are for sleeping on, while weights are for lifting. An inflatable mattress is filled with an air pump. Water weights are filled in a sink. However, these products are analogous, because they preserve a common set of relationships. Mattresses are hard to travel with because they are heavy, so an inflatable mattress removes the heavy component (stuffing) and replaces it with a resource (air) that provides the same functionality and is available at the location where the mattress is to be set up. Likewise, weights are hard to travel with because they are heavy. The water weights replace the heavy component (metal mass) with a resource (water) that provides the same functionality and is available at the location where the weights are to be used.

People have a remarkable ability to notice the similarities between domains that are not alike on the surface. Analogies also allow people to extend their knowledge of one domain by virtue of its similarity to another (Clement and Gentner, 1991; Markman, 1997). This ability to make analogical inferences is crucial to analogical problem solving. When solving a new problem, the problem statement is only a partial match to the known solutions. The key to analogical problem solving is to find known problems that have the same structure as the problem being solved. The solution to the known problem is then a candidate to be applied to the new situation.

For example, if someone were trying to create a set of weights that could be used during travel, then the representation of that domain would not contain any information about potential solutions. Solving this problem by analogy requires matching the problem statement and the obstacles to problem solution (create lighter weights that can be used for travel, because regular weights are heavy) against known problems that have solutions (such as the inflatable mattress). The problem solver can then try to adapt the solution to the new domain.

Adapting solutions is itself not a trivial process (Alterman, 1988; Greiner, 1988). For example, to adapt the inflatable mattress solution to water weights, the weight must be reconceptualized as a container that can be deflated. Then, rather than filling the weights with air, they must be filled with something easily available and heavy; so water is substituted for air. This process of adapting a solution may itself be accomplished by drawing on further analogies.

Of course, solving a new problem by analogy is relatively straightforward if the analogous domain is already present. Designers solving problems that require innovative solutions are in a situation in which they have a problem statement, but the solution is not known. Therefore, they must retrieve domains that are potentially analogous to see whether they know about solutions that can be adapted to solve the problem.

It turns out that retrieving analogous problems is not nearly so easy as recognizing an analogy between two domains that are already being compared. A classic demonstration of this point came from research by Gick and Holyoak (1980, 1983) who told people a story about a general who split his army into groups and had them attack a fortress from a variety of directions, because the main road leading to the fortress had been mined. Later, people tried to solve Duncker's (1945) radiation problem in which they must treat a patient with an inoperable tumor with radiation. Radiation strong enough to kill the tumor will also kill the healthy tissue surrounding it. The solution is to split the radiation into weaker rays and converge them on the tumor so that the tumor is the only area of tissue that receives enough radiation to be destroyed. Despite having seen the story about the general earlier in the experiment, few people in this study recognized the similarity between this story and the radiation problem.

More generally, research on analogical retrieval suggests that people tend to retrieve information from the same domain as the current situation rather than information that is analogous to the current situation (Catrambone, 2002; Gentner, Rattermann, and Forbus, 1993). For example, when solving a problem about oncology (as in Duncker's radiation problem), people are likely to think of other medical solutions, or perhaps other solutions involving radiation. Military solutions are unlikely to come to mind.

This bias to retrieve information on the basis of the overall similarity of the situations arises because most of the time information from the domain currently being encountered is the information that is most useful. For example, generally speaking we want doctors to be reminded of medical situations when treating patients, because that is the information that is typically useful. That is, problem solving is most often successful when working within a domain of expertise.

Cases that require innovation, however, are those for which obvious domain knowledge is not helpful for solving the problem. In these cases, it would be advantageous for experts to be reminded of cases that are analogically similar to the problem being solved. Thus, it is useful to consider factors that improve people's chances to retrieve analogies.

Improving Analogical Retrieval

Some research has explored factors that affect people's ability to retrieve prior knowledge on the basis of analogical similarity. Given the importance of this topic to analogical reasoning more generally, though, it is surprising that more work has not been devoted to this topic.

One set of studies by Clement, Mawby, and Giles (1994) demonstrated that the way that relations are described in the base and target domains affects analogical retrieval. Participants were asked to read short passages; then they read other passages a few days later. The passages presented later told similar stories using quite different characters. For example, one passage might be about alien creatures and a second might be about satellites. Thus, the analogous passages were different in their surface information. Given this setup, people were better able to retrieve the earlier passage they read when the relations were described using general language than when they were described using specific language. For example, the specific relations might involve the alien creatures eating rocks and the satellites taking photographs, while the more general relational language might refer to gathering rocks and collecting photographs. These more domain-general relations made it easier for people to retrieve stories with relational similarities despite the difference in story domains. Thus, the studies by Clement et al. (1994) demonstrate that the content of the stories has an important influence on the likelihood of analogical retrieval.

Nearly all research on analogical reasoning has used written materials. Often, participants have to read relational passages, and then some time later they read a second passage and are asked to recall something similar. Markman, Taylor, and Gentner (2007) examined whether the use of written materials affects analogical retrieval. In their study, they presented people with a sequence of proverbs. For half of the subjects, the proverbs were presented in written form, and for the other half they were read aloud. After each presentation, subjects were either asked to define the proverb or to recall any proverbs they heard previously on the list that were similar to it, either by having similar objects or by having the same meaning.

Some of the sets of proverbs in the list had similar objects and different meanings. Others had dissimilar objects but similar meanings. For example, the proverb, "The swiftest steed can stumble," shares similar objects with the proverb “A rough steed needs a rough bridle,” but is relationally similar to the proverb “The greatest master is wrong from time to time.”

Recall of proverbs that had similar objects in them was good regardless of whether the proverbs were presented in written or spoken form. Interestingly, though, recall of proverbs was much better when the proverbs were presented in spoken form than when they were presented in written form. This finding suggests that previous research may have underestimated people’s analogical retrieval abilities by focusing on written presentation of materials.

Recent research has begun to examine analogical retrieval abilities in the context of innovative design. In one set of studies, mechanical engineering students with some background in design were given descriptions of products that would be useful in later design projects (Linsey, Laux, Clauss, Wood, and Markman, 2007). For example, the participants might read about an inflatable mattress like the one described earlier. Later, they were given difficult design problems (e.g., design weights that can be used for travel). Linsey et al. varied the level of abstractness of the description of both the base analogy in memory as well as the abstractness of the description of the design problem. For example, the air mattress could be described as being filled with a substance at the location where it will be used (a domain-general description) or as being inflated with air in the home where it will be slept on (a domain-specific description).

The results of this study supported the previous work by Clement et al. (1994) in that the domain-general description of the analogous solution was more likely to be used than was the domain-specific description of the analogous solution. Interestingly, people were much better at solving the new problem using the analogous solution when the new problem was described in domain-specific language than when it was described in domain-general language. Second, people in all conditions including the domain-general problem statements, produced numerous additional solutions that were not based on the presented analogous solution. Most of these solutions did not meet the constraints of the design problem. This result indicates that different representations of the design problem facilitate the retrieval of various solutions; therefore, multiple representations of the design problem should be used. Finally, the use of analogy in this study was implicit to some degree. Participants often used the solution from the analogy they were exposed to without explicitly recalling the prior instance. This finding is similar to a conclusion drawn by Christensen and Schunn (2007), who also found that the use of prior solutions by designers was not always accompanied by an explicit recognition of where that solution came from.

However, the structure of this study also suggests another factor that may be important for improving analogical retrieval. Participants in this study solved the problem in phases. In a late phase of the study, designers were given a *function structure* that described the problem. Function structures are representations drawn by designers to convey the abstract structure of a design (Otto and Wood, 2001). Function structures do involve some process choices about the design, but they are more abstract than most descriptions of a design problem are likely to be. Giving participants a function structure that is consistent with a solution to a problem suggested by an analogy also increased people's likelihood of finding the analogous solution. This study did not ask participants to draw their own function structure, so it does not address the question of what kinds of function structures people would draw on their own, given a particular problem statement. Function structures are one type of functional representation within engineering and one of a multitude of design representations. This study indicates that engineers should redescribe their design problems in a multitude of representations, and that other representations of the design problem are likely candidates for facilitating the innovation.

To summarize, previous work on analogical retrieval suggests that representing information in memory using domain-general language yields better analogical retrieval than does representing that information using domain-specific language. This research showed mixed results on the importance of having the current situation represented in domain-general language. Work on design suggests that starting with a domain-specific description of the problem to be solved may be quite useful for solving problems by analogy. Finally, research suggests that significant work by design teams should be done orally, because this modality leads to better analogical retrieval than does written presentation and processing of information.

Tools for Analogical Innovation

This analysis of the role of analogy in innovation suggests that there are two fundamental limitations on the ability of a design team to use analogy to solve a new problem. First, the team is limited to the knowledge possessed by its members. Second, even if a relevant analogous solution is within the knowledge base of its members, the people with that knowledge may fail to retrieve it. Thus, tools for analogical innovation must address these two limitations.

At the outset of this discussion, it is important to note that we have just shifted our terminology from talking about designers to talking about design teams. In practical settings in industry, marketing, and consumer products design, innovation is typically done by teams rather than by individuals. Thus, the composition of the design team plays a crucial role in the success of the innovation team.

This issue is particularly important in the context of analogical problem-solving. Often, design teams are constructed by making guesses about the relevant expertise for that team. For example, a team may have an expert in customer research who has done empirical work on customer needs. The team may also have experts in the particular area of expertise required to create the product (e.g., mechanical engineering, chemistry, or software design). In addition, the team may have representatives from management and marketing. Obviously, there are many difficult issues just in getting teams like this to work together effectively.

Another potential problem is that the analogy necessary to create an innovative solution to a problem may not exist in the heads of this group. Design teams are set up at the start of the innovation process based on the domains known to be relevant to the problem. Obviously, it is not possible to foresee the domains for which there are analogous solutions. Consequently, experts in domains that have potentially innovative solutions to a new problem may not be represented on design teams. This analysis suggests that when a design team is created, individuals with expertise outside of the obvious areas might also be included to provide a perspective on other potential solutions to a problem that might not be obvious to those within the domains of expertise in which the problem is set. In the following two sections, we discuss tools that will be useful for helping design teams maximize their ability to retrieve analogous solutions they know and for extending their knowledge using other potential sources of information.

Tools for Helping Groups to Retrieve Analogies

Helping design teams retrieve analogous problems from other domains requires providing them with techniques for formulating problem descriptions that will capture the relational essence of a problem. That is, the members of a group enter the design session with whatever knowledge they already have, so there is no way to change their knowledge bases. The main element of the problem that can be varied, then, is the way it is being represented by the group.

The default representation that a group is going to have for the problem is quite domain-specific. Design teams are usually very familiar with the problem domain, and quite used to solving problems within that domain. Thus, they have been rewarded in the past for using domain-specific representations of problems. Furthermore, most people have not learned much about human memory. Thus, their attempts to control what they remember are haphazard.

Thus, tools for supporting analogical retrieval must make the innovation process systematic, and must focus on the elements of problem representations most likely to lead to successful analogical retrieval. We address these two issues in sequence. At the outset, though, we should note that the advice to "be systematic" seems antithetical to the lay stereotypes about brainstorming, in which rules and constraints should be thrown out. It is well known, though, that brainstorming sessions often get focused on a small number of ideas that are presented first (e.g., Mullen, Johnson, and Salas, 1991). Furthermore, because people do not know about the factors that affect their memory, a systematic approach can teach people effective methods for generating problem representations that will maximize the chance that the group will find analogous domains.

Generating a problem representation

To systematize the innovation process, tools must provide a scaffold for creating a problem representation that will support finding analogies. The first element of this scaffold is to ensure that designers have a clear understanding and representation of the problem to be solved. This element of problem solving might seem trivial, but there are two reasons why it is not.

First, research on causal reasoning suggests that people often have less causal knowledge about situations than they believe they do (Rosenblit and Keil, 2002). That is, people often believe that they can give a causal explanations for more things than they are actually capable of explaining. Rosenblit and Keil call this *the illusion of explanatory depth*. Similarly, designers may believe that they have a better understanding of the design problem to be solved than they actually do. Thus, as a first step, designers should be encouraged to restate the problem being solved as explicitly as possible. This step will root out aspects of the problem that are actually unclear.

Second, many design teams are confronted with problem statements that are initially vague. For example, Jansson and Smith (1991) gave design teams the simple problem of designing a spill-proof coffee cup. Even in real design settings, the problem statement is often not much more specific than that. Therefore, the members of a design team may not even agree about what problem they are trying to solve.

At the outset of the design process, it is useful for members of the design team to consider a variety of problems that they might be solving. However, as the design process progresses, the team must begin to agree on the problem being solved. Thus, the scaffolding created by innovation tools should begin by ensuring that each member of the design team is trying to solve a specific problem. The members of the group should then share their statements of the problem in order to find points of divergence. The group may choose to consider a variety of different potential problem statements, but it is important that the group agree on the set of problems that they are solving.

What sort of tool would effectively serve to support the development of a problem statement? In innovation contexts, we are used to thinking about computer-based tools. Indeed, many of the chapters in this volume have focused on ways of using the computer and the Internet to support innovation. In this spirit, many of the other tools we discuss in this chapter are best developed in software.

In the case of problem statements, however, the key tool is a systematic structure for creating problem statements. Any mode of description that is typically used by design teams is sufficient to implement this structure. Designers should be encouraged to write out their problem statement. However, they should also sketch when needed (see Tversky and Suwa, this volume) or use formal representations from their domain (such as the function structures discussed earlier).

The problem statement process begins with each member of the design team being given the problem statement as it has been formulated initially. Each team member is then encouraged to provide a detailed description of the problem on their own. They should be explicit about where the problem lies (particularly when redesigning existing products), and what methods are to be brought to bear on solving this problem. Designers must also be explicit about the constraints on the problem to be solved. For example, there are often cost or energy constraints on solutions. If those constraints are not made explicit initially, then teams may develop innovative solutions that cannot be made practical, because they violate fundamental constraints on the problem.

Before settling on a problem statement, the group should also evaluate the degree to which that problem statement is focused on existing solutions for this problem (if any). For example, a company that made film-based cameras in the 1990s might have wanted to make film less expensive to produce. Such an endeavor would probably have focused on the chemicals in film that led to the expense and a search for alternatives. An alternative, however, is to focus on whether there are less expensive mechanisms for capturing images. This formulation of the problem might allow a design group to consider alternatives to image storage beyond film.

To summarize, tools that support the development of good problem statements must focus primarily on the process of generating statements. The purpose of this scaffolding is to ensure that the design team is explicit about the problem being solved, that the design team agrees on the problem to be solved, and that the team considers the possibility that their design problem is too strongly focused on resolving problems with existing solutions to the problem.

Optimizing a problem representation for retrieval

Once a design team agrees on the problem to be solved, one element of the problem-solving process is finding similar problems with known solutions. Because the design team will have domain experts from the problem domain, it is likely that the team will be familiar with many solutions from within that domain. Thus, an important focus of tool development must be for supporting the retrieval of analogous problems.

Tools for Retrieving Analogous Problems from a Broader Knowledge Base

Human memory retrieval is effortless. That is, humans are designed so that information that is active serves as a partial pattern that is completed by other information from memory. A core principle of memory retrieval is *encoding specificity*, which states that information will be retrieved from memory to the degree that the context (e.g., the perceptual and conceptual information that is active) at the time of retrieval is similar to the context at the time the information was put into memory (Tulving and Thomson, 1973). This retrieval occurs automatically.

So what can be done to help people retrieve information? The knowledge people have has already been learned, so there is no way to influence that. Thus, the only lever at the disposal of the tool designer is to influence the representation of the current problem in a way that will be most likely to help group members retrieve prior problems that are analogous to the current problem.

Based on the research reviewed earlier, there are three things that designers must be encouraged to do in order to maximize the likelihood of retrieving analogous problems. First, they must be encouraged to focus on the causal and relational aspects of the problem rather than the superficial contextual elements of the problem. For example, a design team thinking about how to make photographic film less expensive could begin by thinking about improving image-storage media rather than film itself. By recasting the terms of the problem as image storage, the designers can then be reminded of many different methods for storing images (including photocopies and digital scanning).

A second, related aspect of this tool is that it should encourage the use of abstract relational terms to describe the problem. Many of the relational terms we use to describe problems are verbs and gerunds (nouns derived from verbs). Often in technical situations, we use quite precise language. For example, when describing film, we may refer to particular chemical reactions brought about by exposure of chemicals to light. In the previous paragraph, however, photographic film was described as an image storage medium. Storage is a more abstract relation than a description of a chemical reaction.

There are many good tools that can be used to promote more abstract redescriptions of problems. For example, Ward (this volume) talks about online language databases like WordNet (e.g., Miller and Fellbaum, 1991). These databases can be used to find more abstract terms to describe a problem, which are useful for analogical retrieval. Tools for supporting the retrieval process should not simply point people toward these databases, but should also build engines that suggest ways of redescribing problems, given an initial problem description created by a user.

Finally, there is a tendency for tools that support innovation to present information in written format. Because there is some reason to believe that analogical retrieval is easier when information is presented in other modalities, the tools for innovation should encourage discussion during design sessions to make the conditions more conducive to analogy finding.

Tools for Extending the Group Knowledge Base

In many cases of innovation, the relevant domains for solving a problem may not be known by the members of the design team. In this case, it would be useful to have tools to search broader databases to find potential solutions to problems. There are many possible sources for solutions, including the patent database and the broader Internet. In addition, many large corporations have potential stores of proprietary knowledge in their system of internal project reports that may reflect untapped institutional knowledge.

Unfortunately, tools for searching databases tend to be focused on finding information that is from the same domain as the query that is given. For example, typing "travel weights" into Google retrieves a number of examples of existing solutions to this problem, but it does not bring up examples of other products that are analogous to travel weights. Similarly, typing words from the problem statement into the patent database yields other solutions to problems that share surface similarity with the query.

Thus, in many ways, tools for searching electronic databases face a similar problem to that of tools designed to create good retrieval cues for human members of design teams. One problem with using current database search programs is that they are often focused on the specific search terms you enter. There are techniques for creating new queries using synonyms of the terms that are entered, but the more variants to a search term that are used, the more hits are returned from the database. Because most of the entries that are returned are not likely to be useful, the problem becomes one of trying to find a needle in an ever-increasing mound of hay.

While there are no existing analogical search engines at this time, there are many promising avenues for future development. First, research on knowledge representation in artificial intelligence has examined several systems for doing common-sense reasoning in real domains (e.g., Barker et al., 2004; Barker, Porter, and Clark, 2001). The goal of this research is to find ways of representing knowledge in a format that will allow a machine to reason about problems and to retrieve prior problems that might be useful in solving new ones. The representation languages used in these programs create abstract systems of relations that can be used to represent similar problems across domains. Systems of this type have been used successfully in limited domains such as answering academic chemistry questions or suggesting courses of military action.

One reason why systems of knowledge representation may be useful for developing analogical search engines is that they contain links among relational concepts that permit these relations to be redescribed. For example, a group working on developing travel weights might focus on the ability of the user to carry the weights. To explore this example, we can examine the concept *carry* in the library of generic concepts developed by Barker, Porter and Clark (2001). In this library, the specific act of carrying is broken into the concepts of *moving*, *locomoting*, and *holding*. Thus, an innovation team could focus on making any of these steps easier to carry out. Furthermore, queries to a database could focus on novel methods for carrying out these functions. These kinds of knowledge bases have not yet been applied to innovation tools, but they are a promising direction for future research.

One avenue for developing analogy search tools utilizes the concept of *functional decomposition* to redescribe the design problem or concepts. The problem can be transcribed into a set of high-level, domain-independent terms that address the issue of solution domain fixation by eliminating causal and structural relationships that are specific to the current process for solving a problem. So, for example, the concept of *filtering* can be abstracted to *separating*. Furthermore, a common, finite set of terms can be used to represent all (or at least the majority of) problems and domains. One example of such a taxonomy is the Functional Basis (Hirtz, Stone, McAdams, Szykman, and Wood, 2002; Stone and Wood, 2000). This hierarchical set of terms consists of general classes of functions. Previous work illustrates that applying metrics of functional similarity across problem descriptions and potential solutions can identify innovative analogous solutions (McAdams and Wood, 2000).

The major limitation to applying this or another representational scheme directly to search query generation is the limited vocabulary and potential terminology conflict with the database or knowledge repository. The solution to this problem is to apply knowledge extraction techniques to the target repository to augment the representational space with semantically related terms that are specific either to a particular content domain or to a document repository. In this way, the gap between domain-independent, abstract problem descriptions and domain-specific, concrete solutions can be bridged. This methodology has been applied successfully to the functional basis utilizing the U.S. Patent Database as the target search repository. The taxonomy grew from 225 terms to approximately 850 terms to encompass the extent of the patent solutions space, but still represents a finite and tractable information-retrieval problem.

A problem for generating abstract searches is that more-general search terms will also tend to yield more documents. Thus, it is important to develop methods for filtering the set of documents retrieved to focus on those that are likely to be relevant to the current problem. One promising filtering technique is derived from the semantic relationships among terms in the query. For example, documents can be filtered based on whether they contain other terms that frequently co-occur with the terms in the search query. There are many techniques that focus on such co-occurrence relationships to derive semantic similarities among terms (e.g., Landauer and Dumais, 1997).

The final layer of this analogy search methodology uses clustering techniques to manage the (potentially) vast amount of information retrieved from a search. The patent database is especially well suited for domain categorization, because patents exist in well-defined structured classes. Domain clustering promotes both implicit and explicit analogy mapping. Explicit mapping is enhanced by presenting potential analogous solutions in finite and cognitively manageable sets of solutions so that a single innovative solution can be extracted from a smaller solution space. Implicit mapping is enhanced by presenting the user with alternative solution domains that may not have been considered, but where solutions already exist. Increasing awareness of alternative domains can lead to innovative analogous solutions without requiring the user to know about these other patents already.

Summary and Conclusions

There are many roads to innovation. As discussed in this chapter, though, there is good reason to believe that many innovations arise from the reuse of known solutions in new problem domains. People's analogical reasoning abilities are central to finding and applying existing knowledge.

As we discussed, though, it has long been known that analogies are important for solving new and seemingly intractable problems (Polya, 1945). The difficulty lies in finding the relevant analogous domain. In some sense, the analogy between the domains is only obvious in retrospect. At the time that the problem is still unsolved, it may not even be clear what kind of problem is being solved. Thus, finding the relevant domain requires recognizing the nature of the problem being solved.

The tools for innovation that we discuss in this chapter are largely focused on giving people systematic methods for representing the problem that they are solving and for generating cues that they can use for their own memories as well as for retrieval from electronic databases. These aspects of innovation are an under-appreciated part of developing novel solutions to problems. The kind of scaffolding we suggest here will provide innovation teams with the best chance to succeed in their goal to find analogous domains that can be used to solve new problems.

The work discussed here also opens up new avenues for future research. The chapters collected in this volume represent the leading edge of research collaborations among core disciplines involved in innovation. Analogical reasoning in particular will benefit from these collaborations. Most research on analogical reasoning in psychology and computer science has focused on domain-general aspects of analogy such as structural factors that influence the information from one domain that is mapped to another or the relative contributions of surface and structural information in analogical retrieval.

Little research on analogical reasoning focuses on the content of the representations in the base and the target. Basic research in psychology tends to ignore content, because it assumes that it is studying general mechanisms. However, understanding complex reasoning processes like those involved in innovation will require that research attend to the content of the domains being reasoned about. Collaborations between psychologists and engineers (as in the projects described here) are particularly good for this exploration. With this collaboration, the development of new design methods of practical utility for engineers will also lead to new insights into the psychology of basic reasoning processes.

References

ALTERMAN, R. (1988). Adaptive planning. *Cognitive Science, 12*, 393–421.

BARKER, K., CHAUDHRI, V. K., CHAW, S. Y., CLARK, P. E., FAN, J., ISRAEL, D., et al. (2004). *A question-answering system for AP chemistry assessing KR&R technologies.* Paper presented at the The Ninth International Conference on the Principles of Knowledge Representation and Reasoning (KR2004), Whistler, Canada.

BARKER, K., PORTER, B., and CLARK, P. E. (2001). *A library of generic concepts for composing knowledge bases.* Paper presented at the International Conference on Knowledge Capture, Victoria, B.C., Canada.

BASALLA, G. (1988). *The evolution of technology.* Cambridge, U.K.: Cambridge University Press.

CATRAMBONE, R. (2002). The effects of surface and structural feature matches on the access of story analogs. *Journal of Experimental Psychology: Learning, Memory, and Cognition, 28*(2), 318–334.

CHRISTENSEN, B. T., and SCHUNN, C. D. (2007). The relationship of analogical distance to analogical function and pre-inventive structure: The case of engineering design. *Memory and Cognition, 35*(1), 29–38.

CLEMENT, C. A., and GENTNER, D. (1991). Systematicity as a selection constraint in analogical mapping. *Cognitive Science, 15*, 89–132.

CLEMENT, C. A., MAWBY, R., and GILES, D. E. (1994). The effects of manifest relational similarity on analog retrieval. *Journal of Memory and Language, 33*, 396–420.

Duncker, K. (1945). On problem solving. *Psychological Monographs, 58.*

Gentner, D. (1983). Structure-mapping: A theoretical framework for analogy. *Cognitive Science, 7,* 155–170.

Gentner, D., Holyoak, K. J., and Kokinov, B. (Eds.) (2001). *The analogical mind.* Cambridge, Mass.: MIT Press.

Gentner, D., Rattermann, M. J., and Forbus, K. D. (1993). The roles of similarity in transfer: Separating retrievability from inferential soundness. *Cognitive Psychology, 25*(4), 524–575.

Gick, M. L., and Holyoak, K. J. (1980). Analogical problem solving. *Cognitive Psychology, 12,* 306–355.

Gick, M. L., and Holyoak, K. J. (1983). Schema induction and analogical transfer. *Cognitive Psychology, 15*(1), 1–38.

Greiner, R. (1988). Learning by understanding analogies. *Artificial Intelligence, 35*(1), 81–125.

Hirtz, J., Stone, R. B., McAdams, D. A., Szykman, S., and Wood, K. L. (2002). A functional basis for engineering design: Reconciling and evolving previous efforts. *Research in Engineering Design, 13,* 65–82.

Holyoak, K. J., and Thagard, P. (1995). *Mental Leaps: Analogy in creative thought.* Cambridge, Mass.: MIT Press.

Jansson, D. G., and Smith, S. M. (1991). Design fixation. *Design Studies, 12*(1), 3–11.

Kolodner, J. (1993). *Case-based reasoning.* San Mateo, Calif.: Morgan Kaufmann Publishers.

Landauer, T. K., and Dumais, S. T. (1997). A solution to Plato's problem: The Latent Semantic Analysis theory of acquisition, induction, and representation of knowledge. *Psychological Review, 104*(2), 211–240.

Linsey, J. S., Laux, J., Clauss, E. F., Wood, K. L., and Markman, A. B. (2007). *Effects of analogous product representation on future design-by-analogy.* Paper presented at the International Conference on Engineering Design, ICED'07, Paris, France.

Markman, A. B. (1997). Constraints on analogical inference. *Cognitive Science, 21*(4), 373–418.

Markman, A. B., Taylor, E., and Gentner, D. (2007). Auditory presentation leads to better analogical retrieval than written presentation. *Psychonomic Bulletin and Review, 14*(6), 1101–1106.

McAdams, D. A., and Wood, K. L. (2000). *Quantitative measures for design by analogy.* Paper presented at the Proceedings of the 2000 ASME Design Theory and Methodology Conference, Baltimore, Md.

Miller, G. A., and Fellbaum, C. (1991). Semantic networks of English. *Cognition, 41,* 197–229.

Mullen, B., Johnson, C., and Salas, E. (1991). Productivity loss in brainstorming groups: A meta-analytic integration. *Basic and Applied Social Psychology, 12*(1), 3–23.

Newell, A. (1990). *Unified theories of cognition.* Cambridge, Mass.: Harvard University Press.

Newell, A., and Simon, H. A. (1963). GPS: A program that simulates human thought. In E. A. Feigenbaum and J. Feldman (Eds.), *Computers and thought.* R. Oldenbourg KG.

Otto, K., and Wood, K. L. (2001). *Product design: Techniques in reverse engineering and new product development.* Upper Saddle River, N.J.: Prentice Hall.

Polya, G. (1945). *How to solve it.* Princeton, N.J.: Princeton University Press.

Rosenblit, L., and Keil, F. C. (2002). The misunderstood limits of folk science: An illusion of explanatory depth. *Cognitive Science, 26*, 521–562.

Stone, R. B., and Wood, K. L. (2000). Development of a functional basis for design. *Journal of Mechanical Design, 122*, 359–370.

Tulving, E., and Thomson, D. M. (1973). Encoding specificity and retrieval processes in episodic memory. *Psychological Review, 80*, 352–373.

CHAPTER 6

CONSTRAINTS AND CONSUMER CREATIVITY

C. PAGE MOREAU

DARREN W. DAHL

THE ubiquitous phrase "thinking outside of the box" implies that creative thought requires breaking through the walls that constrain ideas. Indeed, it is much easier and more cognitively efficient to solve problems by retrieving known, established solutions (i.e., paths of least resistance: see Finke, Ward, and Smith, 1992; Ward, 1994). Deviations from known solution paths can require significant time and cognitive effort (Perkins, 1981). What the conditions are that force people from retrieving well-established solution paths and the implications of these deviations for both the outcome of a creative task and the person's experience during it are the focus of this chapter.

While the importance of constraints in creative tasks has been noted by researchers in psychology (e.g., Costello and Keane, 2000; Finke et al., 1992; Stokes, 2001), few studies have examined how constraints influence individuals' cognitive processes, their subjective experiences, and the outcomes produced in these situations. An investigation of all of these aspects of creativity is especially critical in a consumer context as manufacturers and retailers vie to develop and sell products that satisfy consumers' apparent demand for creative experiences and unique outcomes.

Constraints and Outcome Creativity

A unique aspect of many creative tasks lies in the fact that the problem or challenge itself is often not well defined (Guilford, 1950; Newell and Simon, 1972). In contrast to studies of choice in which problem representations are defined by the task (Bettman, Luce, and Payne, 1998, p. 208), an infinite number of satisfactory solutions may exist for a creative challenge, depending upon how the individual constructs the task representation. Furthermore, as in any constructive process, a number of environmental or individual factors may constrain the structure and content of the plan.

To understand how constraints influence cognitive thought in a creative task, we used the Geneplore model (Finke et al., 1992) as a theoretical basis and ran a series of experiments. The Geneplore model describes the two key cognitive inputs for creativity: generative and exploratory processes. This model has the advantage of distinguishing between the cognitive processes used in creative cognition and the mental structures upon which they operate (Ward, 2001). Thus, creative and non-creative thinking can be conceptualized along a continuum with no absolute boundary separating the two. It is the extent to which both generative and exploratory cognitive processes are utilized in developing a solution that determines the likelihood that a more creative idea or product will result (Moreau and Dahl, 2005; Ward, 2001).

Generative and Exploratory Processes

In the initial stages of a creative task, generative processes are used to create preliminary mental representations of a solution, called "preinventive structures," that serve as a precursor to the final creative product (Finke et al., 1992, p.19). The generative processes used to construct these representations have received extensive attention in the literature: the retrieval of existing structures from memory (e.g., Perkins, 1981), the creation of associations or combinations among the retrieved structures (e.g., Murphy, 1988), and analogical transfer from one domain to another (e.g., Gentner, 1989). In contrast to generative processes, exploratory processes search for meanings to attach to and/or to interpret the preinventive structures. One basic approach to interpretation is to search for potential functions. Other exploratory processes that are often used to attach meaning to these novel forms include evaluating the structure(s) from different contexts or perspectives, interpreting it as a possible solution to a salient problem, and/or searching for practical or conceptual limitations suggested by the structure's form.

If the exploratory processes yield a satisfactory interpretation of the preinventive structure, the path to a creative product or idea is quite short. Alternatively, the exploratory processes may not be completely successful. Typically, the creative process is one of cycling between generation and exploration, with the initial representation altered and updated with each cycle until a satisfactory final product or solution is achieved (Finke et al., 1992).

The Influence of Input Constraints on Generative and Exploratory Processes

While the Geneplore model describes the requisite cognitive processes involved in creativity, earlier empirical tests of the model had been realized by judging the *outcomes* generated in different contexts and inferring from the outcome that the mechanisms theorized actually occurred. Thus, we developed a set of studies to understand the effect of constraints on participants' information-processing strategies when engaged in a creative task (Moreau and Dahl, 2005). In these studies, we asked participants to design a toy for a child (aged 5–11), while manipulating the constraints on the inputs to be used in the design. All participants were shown 20 different shapes (see Fig. 6–1).

In the first study, half of the participants were able to select five of the shapes, while the other half were randomly assigned a set of five. Following this first constraint manipulation, the second input constraint was manipulated by telling participants they either must use all five shapes in their toy design or that they could use as many shapes as they would like. Once the manipulations were complete, participants were given an unlimited amount of time to design, draw, and name their toy idea. After completing their designs, participants described the process they had used to come up with the toy idea. These written protocols were coded by research assistants in order to assess the degree to which generative and exploratory processes were used during the design task.

To assess the creativity of the toy designs, three senior design professionals were used as expert judges. Each judge rated each design on six seven-point scales designed to capture the novelty (three items) and appropriateness (three items) of the toys. The items were summed to form two indices: one for novelty and one for appropriateness.

Moreau and Dahl (2005; Study 1) demonstrated that when multiple constraints were imposed during a creative task, more generative and

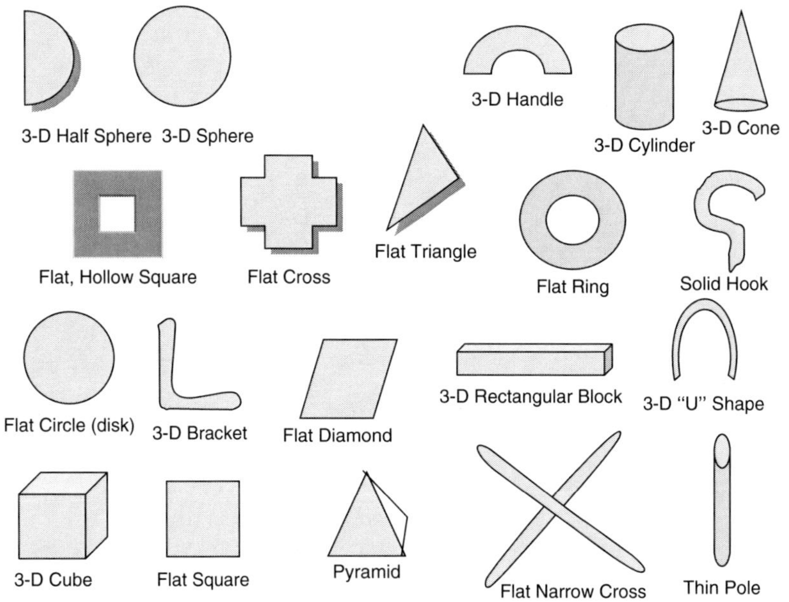

Figure 6–1 Choice of shapes used in the studies. Reprinted with permission from Moreau, C. Page and Darren W. Dahl (2005), "Designing the Solution: The Impact of Constraints on Consumer Creativity," *Journal of Consumer Research*, 32 (June), 13–22.

exploratory processes were employed. Specifically, the data demonstrated that only when both constraints were active (the five shapes were assigned and all five had to be used in the toy design) were participants forced off a "path of least resistance" (Ward, 1994). For example, one participant in this condition was assigned a cone, a cylinder, a disk, a thin pole, and a flat ring. From this assortment, he created a "rain bath" toy in which the cone and the pole were used to create an umbrella of sorts to deflect water into a broader cylindrical bath at the bottom (formed by the disk, the cylinder and the flat ring). The participant described the toy: "When the kids go in it to play, it seems like they are surrounded by a waterfall."

Following a path of least resistance requires the retrieval and implementation of a known solution (Ward, 1994), and the constraints appeared to work together to impede participants' ability to do that. The results also revealed a positive correlation between the judged novelty of the toy and the degree to which the participant used generative and exploratory processing. No relationship, however, was identified between appropriateness and the degree of generative and exploratory processing.

By employing process-tracing methods, this study established an empirical link between the generative and exploratory processes used during a creative task and the actual judged novelty of the outcome. While this link was hypothesized by the Geneplore model (Finke et al., 1992), there had been little empirical evidence to confirm its existence. This study also demonstrated that the path of least resistance is a strong force requiring that more than one constraint be active in order to pull a participant away from the default processing strategy. When the participants had some degree of choice, either in selecting the five components themselves or in selecting among the established set of five, they did not stray far from the path of least resistance. Only in the extremely constrained situation, when all of their input choices were removed, did the participants engage in extensive generative and exploratory processes.

A follow-up study was used to support the claim that participants who selected their inputs were following a path of least resistance (Moreau and Dahl, 2005; Study 2). The exact same methodology was used in this second study, with one exception: input selection or assignment occurred *before* participants were told that the creative task was "to make a toy." Thus, participants who were able to choose their five shapes could use any number of different decision rules to guide their selection, but the rule chosen would probably be irrelevant to the task at hand. Therefore, no path of least resistance would be available to follow. The expectation, then, was that no differences would emerge between those who selected their own shapes and those for whom the shapes were assigned. Effectively both groups should have been equally constrained by this manipulation, leaving a prediction of only a main effect of the second constraint. Specifically, those who were forced to use all five shapes were expected to show greater evidence of generative and exploratory processes than those who were able to use as many of those five as they would like. The results were consistent with these expectations, thus providing more evidence that the constraints were forcing people off the path of least resistance.

These two studies provided solid evidence that input constraints, when significant enough, can force participants to use more generative and exploratory processing. One reason that people may avoid using this type of processing is that it is both more cognitively effortful and time-consuming than following a path of least resistance. Thus, another study was designed to understand the influence of time constraints on generative and exploratory processing during the creative process (Moreau and Dahl, 2005; Study 3). In this study, half of the participants experienced both input constraints while the other half were unconstrained. The time constraint was manipulated by allowing participants either five or 50 minutes to complete the design task.

Our expectation was that the time constraint would have little effect on the participants for whom inputs were not constrained, since they could retrieve a known toy, choose their shapes, and use as many of the selected shapes as they wished to design the toy. Here, participants could follow a path-of-least-resistance strategy and do so very efficiently, so that the time constraint would have little effect on the extent of the generative or exploratory processing. However, for those subject to the input constraints, the time constraint was expected to significantly diminish the extent of generative and exploratory processing. This expected interaction was supported by the data, providing additional support for the hypothesized mechanisms underlying the relationship between constraints and creative processing.

Constraints and the Creative Experience

The previous set of studies demonstrated that highly constrained participants were more likely to engage in the type of cognitive processes that positively predicted the novelty of the outcome. However, the experiences of these participants may have also differed significantly from those who were subject to fewer input constraints. The previously described study did not examine this aspect of creativity: the subjective experience of the individual engaged in the creative task. Given that individuals often seek out opportunities to be creative in a consumption context, it is important to understand the effects of constraints on their experience and satisfaction with the process. For example, many consumers voluntarily undertake creative tasks when they could more easily accomplish their outcome goals by hiring an expert to solve the problem (e.g., an interior designer, a wardrobe consultant, a travel agent).

There has been little empirical research to suggest the specific motivations underlying consumers' desire to be creative and the effects that constraints may have on these motivations. Thus we found it necessary to draw upon a broader model of motivation to understand consumers in the creative arena. Because self-determination theory is more specific than other motivation theories (e.g., flow theory: Csikszentmihalyi, 1990), it was selected as the theoretical framework for understanding the consumers' subjective experience in creative tasks (Deci and Ryan, 2000). The basic tenet of self-determination theory is that three key needs underlie human motivation: the needs for autonomy, competence, and relatedness, each described further below (Deci and Ryan, 2000). In a recent commentary, Hennessey (2000) advocated the use of self-determination theory in understanding the social psychology of creativity and requested that researchers "think more about how self-determination theory might be specifically applied to the creative process." Self-determination theory posits that "a full understanding not only of goal-directed behavior, but also of psychological development and well-being, cannot be achieved without addressing the needs that give goals their psychological potency" (Deci and Ryan, 2000, p. 228).

The Influence of Input Constraints on Perceived Autonomy, Competence, and Task Enjoyment

Autonomy

The need for autonomy reflects the desire for self-governance, volition, and an individual's wanting to self-organize their experience (Deci and Ryan, 2000; Sheldon and Elliot, 1999). An individual's need for autonomy, however, does not imply a need to behave completely independently of external forces. Rather, autonomy "concerns the extent to which people authentically concur with those forces that do influence their behavior" (Deci and Ryan, 2000, p. 14). As such, the effect of constraints on consumers' perceptions of their autonomy will depend on whether they concur with the presence of the constraints or reject them.

Competence

White (1959) asserted that people are motivated to have an effect on their environment as well as to attain valued outcomes within it, and Deci and Ryan (2000) argue that it is this need for effectance or competence that motivates behavior. Csikszentmihalyi (2000) reinforces this position, noting that many objects are purchased to reinforce self-esteem and competence:

> Many [products] are acquired because they allow the person to practice and perfect a special skill which is important to his or her identity, such as musical instruments, tools, photo equipment, books that reflect the person's interests, sports and gardening equipment, and so forth (p. 269).

How, then, do input constraints similar to those operating in the prior studies impact self-perceptions of competence? Using a similar argument to that used above for autonomy, one can see that it is likely that active constraints will reduce participants' perceptions of competence since the constraints work to move them off of the path of least resistance. When individuals follow a familiar course of action or have the ability to choose the path they wish to follow, they are likely to feel more competent in their pursuits than when they operate in an unfamiliar, forced way.

Relatedness

While autonomy and competence have been found to be the most powerful influences on motivation, relatedness—the desire to belong and feel connected—can also be a critical motivator (Deci and Ryan, 2000; Ryan and Deci, 2000). These researchers, however, acknowledge that "there are situations in which relatedness is less central to intrinsic motivation," such as when "people engage in intrinsically motivated behaviors (e.g., playing solitaire, hiking) in isolation." Certainly, in a creative context, the need for relatedness can play a key role in motivation (e.g., brainstorming sessions, ad copy development—see Johar, Holbrook, and Stern, 2001), but in the context of our research, we simply hold relatedness constant across all conditions and examine the individual's creative process in isolation.

In an initial study using engineering students as participants, we make an attempt at understanding the influence of constraints on participants' perceptions of autonomy and competence (Dahl and Moreau, 2005; unpublished manuscript). In this study, the participants were again asked to design a toy, and the two input constraints were manipulated between the 153 participants. Task enjoyment was measured using six nine-point scale items. On each of the scale items, participants reported the degree to which they enjoyed the creative process. For example, one item asked participants to report the extent to which they agreed with the following statement: "I had a lot of fun creating my new product concept."

Autonomy was measured using four scale items. On each nine-point scale, participants reported the degree to which they felt that they worked independently and had control over their actions during the task. For example, one item asked participants to report the extent to which they agreed with the following statement: "I felt like I had a lot of control during the development process."

A task-specific competence measure was constructed following Deci and Ryan (2000), Csikszentmihalyi (2000), and Fournier and Mick (1999). Csikszentmihalyi (2000) defines the need for self-esteem as a need "to feel competent, respected, and superior," and includes as examples consumption items that "allow the person to practice and perfect a special skill which is important to his or her identity" (p. 269). Our measure of these constructs asked participants to "please tell us how participating in this new concept development task made you feel." The question was followed by a series of nine Likert scale items that asked people to indicate the degree to which they felt smart, competent, talented, inspired, proud, and intelligent.

The results of this experiment revealed that when participants were not allowed to select their shapes, perceptions of both competence and autonomy were significantly lower than when shape selection was their choice. Furthermore, participants who were forced to use all of their parts also reported lower levels of autonomy than those who could use as many as they wanted. However, this manipulation did not significantly influence their perceptions of competence.

What were the effects of the constraints on task enjoyment? Participants who were able to select the parts for their new product enjoyed the task significantly more than those to whom the parts were assigned. This finding is interesting, given that in our earlier study, the effect of this manipulation was to *increase* the perceived creativity of the toy. While the participants for whom the parts were chosen may have produced more objectively creative products, they enjoyed the process less. A similar pattern resulted from the "use all" constraint. Participants who were forced to use all five shapes enjoyed the task significantly less than those who could use as many parts as they wanted. A mediation test was used to determine whether autonomy and competence mediated the effects of the constraints on task enjoyment (Baron and Kenny, 1986). While autonomy and competence were both positive predictors of task enjoyment, neither fully mediated the effects of the constraints.

This study explored the motivations underlying and influencing an individual's subjective experience in a creative task. While constraints have demonstrated their ability to increase generative and exploratory processing, the same constraints in this study served to diminish the individual's subjective experience. Indeed, a constraint on the ability to choose inputs and a lack of freedom in deciding what inputs to include in the design solution were shown to be detrimental to a participant's perceived competence, autonomy, and resulting task enjoyment.

These findings are somewhat surprising given the more general phenomena of retailers (e.g., Michaels, Lowe's, Martha Stewart) that have achieved financial success through offering more constrained creative opportunities for consumers in the marketplace. To better understand this apparent contradiction, we found it important to take a closer look at the types of constraints operating. In the studies described thus far, the constraints employed targeted the inputs used by the participant when constructing their new idea. The constraints did little to restrict the outcome of the task since a great number of solutions could be constructed to create a toy. Furthermore, no explicit process constraints, such as step-by-step instructions, dictated how the creative task should be accomplished. Therefore, we ran an additional set of studies to broaden our understanding of the influence of constraints on a creative task (Dahl and Moreau, 2007). These new studies included situations where the creative outcome itself is constrained and where a normative process for approaching the creative task is suggested.

The Influence of Process and Outcome Constraints on Perceived Autonomy, Competence, and Task Enjoyment

In many creative consumption situations a complete solution representation is provided to the consumer. For example, many of Martha Stewart's kits provide pictures of what the end product (e.g., cookies, pumpkins, or embroidery) *should* look like. The Home Depot Expo provides samples of renovated kitchens and bathrooms from which the do-it-yourselfer can choose. Even many cookbooks provide pictures of the entrees along with the recipes. Thus, in many "creative" situations, a visual representation of the creative outcome is provided *a priori*. As an article in *Forbes* noted, many retailers are effectively selling "somewhat individual" outcomes by offering "self-expression for the time-deprived" (Rossant, 1996, p. 152), saving consumers the cognitive effort of developing a representation of the solution.

Further, it is often the case that creative activities offered to the consumer involve detailed step-by-step instructions to achieve task completion. Examples of these types of creative instructions include assembly guidance in model building, direction in tool and material usage in scrapbooking, and detailed recipe information in cooking and baking activities. These types of normative instructions, while providing guidance and facilitating success in completing a creative task, can constrain the individual to a set course that, if followed, limits the approach the consumer takes in completing the activity.

Situations involving these types of constraints best replicate real-world consumer offerings in which the creative target (e.g., a needlepoint nature picture), the inputs (e.g., the thread colors and canvas), and/or the process itself (e.g., step-by-step instructions for the stitching procedure to achieve the specified nature picture) are dictated to the consumer. For an individual, this situation both provides a targeted goal and facilitates the creative process with a solution path that limits the possible approaches towards the solution outcome. Two experiments were conducted to understand how the provision of these constraints enhances or detracts from perceptions of autonomy, competence, and overall task enjoyment.

In both experiments, participants were brought into the lab in small groups of two to five and were given the task of actually making and baking a cookie (the laboratory was outfitted with work stations [see Fig. 6–2] and convection ovens). In the first experiment, all participants were told the following: "Your workstation includes all of the ingredients and tools you will need to make a cookie. Once you are ready for baking, please signal the experimenter. When the cookie has baked, it will be returned to you for decoration."

The outcome constraint was then placed on half the participants, who were given a picture of a decorated cookie and told to make a copy of it. The remaining participants were simply told that they could make and decorate a cookie any way they would like. Finally, those for whom the process constraint was operating received more detailed step-by-step instructions; the others received no additional information. After finishing the cookie-making process, participants reported task enjoyment, perceived autonomy, and competence.

The results demonstrated that task enjoyment was at its highest when one of the two constraints (process or outcome) was operating. When neither constraint was active, task enjoyment was as low as when both constraints were active. Interestingly, there was no main effect of either constraint.

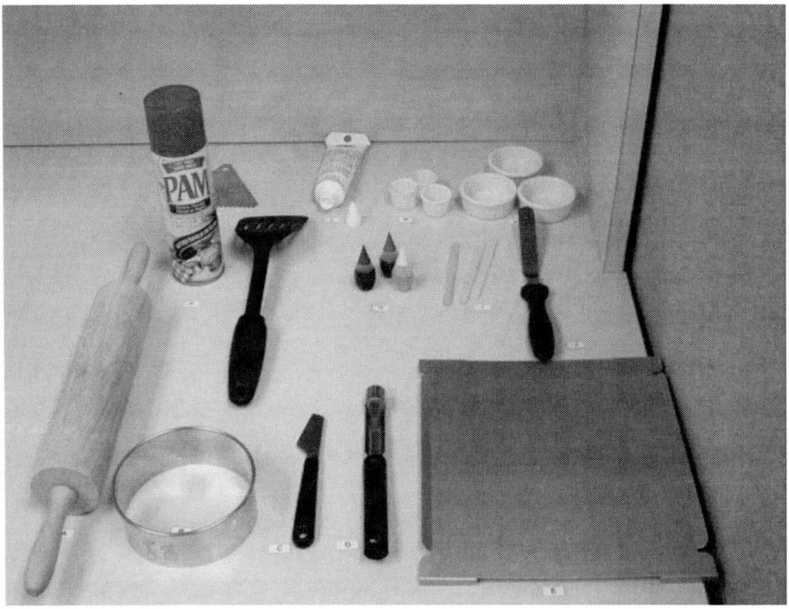

Figure 6–2 The cookie workstation.

Participants experienced an equal level of enjoyment when making their own cookie with instructions as when making the cookie we told them to make without instructions.

Both perceived competence and autonomy were significant, positive predictors of task enjoyment and were both at their highest when participants made and designed their own cookie with the help of step-by-step instructions. Both measures were also high when participants were forced to make the target cookie without instructions. Perceived competence fully mediates the effects of the constraints on enjoyment while perceived autonomy was a partial mediator.

The Influence of Skill and Constraints on Perceived Autonomy, Competence, and Task Enjoyment

Like the paint-by-number kit manufacturers in the 1950s, the initial study assumed no level of prior experience in our study participants. Randomization effectively mitigated any effects that such differences in skill would have on the measures of interest. Yet, prior skill levels are expected to have an important influence, not only on one's likelihood of purchasing a product offering constrained creativity, but also on the likelihood of enjoying the experience offered.

Numerous studies have demonstrated that those with greater skill at a particular task are better able to use their own internal knowledge as a source of guidance than those with lower levels of skill (see Alba and Hutchinson, 1987; Sanbonmatsu, Kardes, and Herr, 1992). For consumers who undertake creative activities as hobbies rather than careers, products offering constrained creativity offer an alternative to this time-consuming and all-encompassing approach. Even in this nonprofessional arena, those with higher skill are also likely to be better equipped to provide their own guidance and feedback than those with less skill. When a task offers them the freedom to follow their own internal guidance, higher-skilled people are likely to enjoy a task more than those with lower skill levels.

In a follow-up study, a similar methodology was employed with the target outcome manipulated between participants and prior baking skill measured using a three-item scale (all participants were given instructions). Competence, autonomy, and task enjoyment were measured using the same scales reported in the earlier studies.

Higher-skilled participants who were allowed to make their own cookie (no target outcome) reported the highest level of task enjoyment, significantly greater than the lower-skilled participants without a target outcome. A similar pattern was observed for perceived competence, and mediation tests confirmed that competence largely explained the pattern of results observed for enjoyment. The two independent factors had significant effects on autonomy, with the provision of the target outcome decreasing it and higher skill levels positively correlated with it.

Taken together, these two laboratory studies demonstrate that constraints on the process and outcomes of creative tasks influence consumers' experiences while undertaking them. Furthermore, the prior experience of the target audience must be considered when making decisions regarding the number and extent of the constraints. While these studies carry the benefits of experimentation, such as the ability to infer causality, they are also subject to the limitations of the methodology. The studies were run with a limited set of operationalizations of the key constructs, thus limiting the ability of the findings to generalize both to different types and levels of constraints and to more diverse creative settings. Further, participants in these studies chose to participate in exchange for course credit, not because of an intrinsic motivation to do so. Since consumers choose to buy creativity kits and undertake creative projects on their own volition, this is also a factor that limits both the studies' theoretical and their managerial implications.

Given that an examination of creative experiences is new to the consumer research, we found it is important to augment the laboratory studies with qualitative data better approximating the motivations and reactions of real-world hobbyists. Consequently, we designed a qualitative study to confirm that competence and autonomy are central motivations for creative people, to examine how these motivations manifest themselves across creative contexts, to identify other sources of motivation underlying creative endeavors, and to provide a basis for stimulating future research in this area. Another crucial goal of the following study was to gain a broader understanding of how the constraints we used in our study (e.g., instructions and target outcomes) are viewed and used by individuals in actual creative contexts.

Broader Motivations for Undertaking Constrained Creative Experiences

We conducted, taped, and transcribed twelve interviews that were conducted with informants representing eight different hobby areas: scrapbooking, modeling, cooking, jewelry-making, card-making, sewing, carpentry, and quilting (see Dahl and Moreau, 2007; Study 1). After each interview, the data were analyzed following a constant comparative technique to ensure that the ensuing interviews would reflect any additional relevant issues that were raised (Corbin and Strauss, 1990). Through joint discussions, we reached consensus and ensured that each factor or theme appeared in the data repeatedly to achieve concept saturation (Glaser and Strauss, 1967; Kirmani and Campbell, 2004; Wallendorf and Belk, 1989).

Table 6–1 lists the seven different motivations that emerged from the data. The first and most frequently mentioned motivations were those of *competence* and *autonomy*, both underlying factors contributing to a consumer's private sense of accomplishment.

Learning how to improve on a skill set also serves as an important impetus for participation in the hobby. Learning opportunities enable the individual to develop a more refined set of skills and techniques. Interestingly, learning often involves a community of fellow crafters, meeting a social need, as discussed later. Additional motivations for *relaxation and engagement* appear to be process-related. For many of the informants, the process was

Table 6-1 Basic Motivations for Undertaking Creative Tasks.

Basic Motivation	Definition	Examples
Competence	Anticipated satisfaction derived from completing a creative project successfully.	"Most of the time I feel really good because you're taking pieces of something and putting it into an actual finished product." "Producing something tangible is a really nice feeling."
Autonomy	Enjoyment derived from the freedom to choose the process and/or design of the creative task.	"You made it yourself; you chose the colors and stuff, so it's customized. It feels like it belongs to you." "I feel happy when I do my models. I feel like I can accomplish something. All the choices you make are your own. You make every decision for yourself." "I make a lot of things you can't find <in stores>. For originality and fabric quality, I prefer to make certain things myself."
Learning	Desire to attain or improve the skills necessary for completing creative projects.	"I like the learning opportunity and sort of look at what the other guy is doing. At meetings, they'll do little presentations on techniques." "I learned through mistakes. The first 5 or 6 car models I bought, I was able to build them but they didn't look as good as the models I build now."
Engagement and Relaxation	Anticipated satisfaction derived from immersion in the creative process itself.	"Sitting in an office all day, coming home, and building something with my hands and taking all my attention is a good way to me to relax and wind down." "It's meditative plus it's a relaxing thing." "There's something about working with wood that's very pleasing to a lot of the senses."
Self-Identity	Desire to reinforce or enhance one's self-perceptions of creativity.	"I think I'm creative and people at work tell me I'm creative because nobody else does stuff like this. I kinda came to realize that I'm different, sort of." "It makes me feel unique because not everybody does this kind of work. People think I'm creative."
Public Sense of Accomplishment	Anticipated satisfaction derived	"It's also for self-gratification when you show it to the person and they ooohhh and

	from others' recognition of one's own creative accomplishments	ahhhh . . . It's gratifying to have someone appreciate it when you give it to them." "The other morning for a contest, I finished a model at 4 a.m. and then the first thing I did was take pictures and post it on the Net for my friends to see." "At weddings, when we give the quilts, our family and friends look at us and admire what we've done and the fact that we're so close." "Well, it's gratifying when you're cooking for a group of people who are really appreciative and they give you a lot of compliments." "My husband really likes it (the scrapbook) and that makes me happy."
Community	Desire to share creative experiences with others who are similarly motivated.	"Everyone's a bit different, not my normal social group, but something like this brings us together . . . we share information and there's a positive feeling in model groups." "We go to one or two meetings a month and it's the same people . . . Even though I don't do stuff with them outside of that. I thought about giving up 'Stamp It Up' because I'm not making money, but I can't give it up. I'd have to give up that whole part of my life."

Reprinted with permission from Dahl and Moreau (2007), Thinking inside the box: Why consumers enjoy constrained creative experiences, *Journal of Marketing Research, 44*(3), 357–369.

both engrossing and relaxing, allowing them to free their minds from other worries. The hobbies also appear to create or reinforce the hobbyist's own *sense of identity*. For many, the hobby allows them to claim creativity as a core characteristic of their personality.

The motivation for a *public sense of accomplishment* also emerged. Informants cited examples of positive feedback from peer hobbyists, appreciation from gift recipients, and admiration from friends and family as important outcomes of the creative process. A hobbyist *community* also provides a forum for public accomplishment, but beyond that, the community provides a set of people with uniquely common interests, the final key motivator for many of our informants. Even though several hobbyists mentioned that they wouldn't normally socialize with the people in their hobbyist groups, they found the companionship invaluable. Examples of each of these motivating factors are provided in Table 6–1.

The interviews also revealed discussions of how creativity products (e.g., kits, how-to guides) affect their private sense of accomplishment. The trade-offs between competence (e.g., instructional guidance) and autonomy (e.g., creative freedom) associated with the creative products were highly salient to the people, dominating the bulk of the discussion on the topic of constraints. Table 6–2 summarizes the pros and the cons, providing multiple examples of each.

Four major advantages were reported for creativity products (e.g., kits, models, patterns, and recipes) which served to improve the hobbyist's sense

Table 6–2 The pros and cons of the constraints imposed by creativity products (kits, models, recipes and patterns).

Pros	Examples
Ease of Use/Efficiency	"I first started when my parents bought me a little model. They're very easy to assemble. Right off the bat, you cut and glue it on. That's it." "There are people who do scratch building [building without the kit]. It is much more difficult. I don't have the time to do that."
Low Skill Requirements	"I think they [novice stampers] like it when they first try it and it looks good . . . even the first time." "It gives you a starting point, and gives you the basics to put something together. Because you know, I'm not a seamstress. I'm not that skilled so it definitely gives me a starting point." "I can see it [the kit] being helpful for, you know, a beginner who doesn't really have the schooling for doing blueprints, and doing cutting lists, and actually figuring out how many sheets of plywood they need to buy." "I wouldn't have the imagination to come up with the different ingredients . . . like a pumpkin spice cream-cheese icing . . . on my own."

Certainty of the Outcome	"I do like to have a picture so I know what it's supposed to look like, so I can see if mine looks as good as the picture does." "If you have a picture and a recipe, you're pretty well guaranteed that it will come out looking like that and taste what you think it's going to taste like." "I'm a big fan of box kits. If you buy it from the box, you try to make it look like the box."
Learning Opportunities	"I learned from the models how to paint. You have to spray fast. There are a lot of skills to painting." "When I was first learning how to sew, that's what you did was follow the pattern, which gives you lots of instructions. I couldn't sew without them."

Cons	Examples
Uniformity of the Outcome	"It is good to have a product that looks exactly like the picture but then it is not special any more, right?" "With kits, there's nothing on your own there. It's just someone else's and all you're doing is assembly. It's like buying at IKEA and saying that you made it." "Kits are like the lazy man's easy way out. Scratch is just the classier thing to do." "I like to customize the kit in order to 'put my own stamp on it.'" "I've outgrown the kits. Sometimes I do use the same pattern, but I try different colors and sometimes the designs look actually quite a lot different." "I definitely like it when I come up with my own idea, just because I don't want to copy something else. I would feel kind of lame. I would rather make it myself and have my own idea and feel like I was creative."
Decrease in Process Enjoyment	"When you're following such strict guidelines, it's pretty frustrating and probably more challenging than when you're just freewheeling. It's pretty constricting." "I think it's less fun if someone is telling me exactly what step to do and what not to do because it's not creative." "I use the pattern as a starting point and go from there. It's like, who wants to follow *anything* exactly by the rules?"

(Continued)

Table 6–2 (Continued)

Cons	Examples
Mismatch Between the Challenge of the Task and the Hobbyist's Skill Level	"Some of the companies have products that are a lot more complicated than they ever have been in the past. They've got all these new technologies that they use and the parts are a lot smaller; they're a lot finer; they're a lot more delicate.... You'll beat yourself over the head over one of these things." "They [the kits] are for those self-proclaimed non-creative people. I see those kits and think 'Oh God.' They're pre-made. You just literally glue it on."

Reprinted with permission from Dahl and Moreau (2007), Thinking inside the box: Why consumers enjoy constrained creative experiences, *Journal of Marketing Research, 44*(3), 357–369.

of competence. The creativity products enabled the informants to complete a creative task with greater ease, and in many cases, the products also allowed people to accomplish their creative goals despite a lack of task-relevant skills. Finally, the products helped consumers develop relevant skills, again reinforcing their sense of competence.

There were three consistent disadvantages to the creativity products noted as well. Informants cited the lack of freedom in the process and lack of uniqueness of the outcome as the primary drawbacks to kits, patterns, models, and recipes (see Table 6–2). Both of these factors limit the consumers' autonomy. Informants also were critical of creativity products that offered tasks that did not match their own skill level. This factor mirrors the ease and efficiency factor and is largely an indicator of perceived competence.

In this qualitative study, individuals' diverse motivations for undertaking creative tasks were established, as well as a better understanding of how constraints influence those experiences. The two motivations discussed most frequently, competence and autonomy, were also the two most affected by the constraints imposed by kits (i.e., instructions and target outcomes). Creativity products across the differing hobbies were shown to provide the needed guidance (and often, raw materials) necessary to competently complete a creative task in a reasonable time. The creativity products also reduced perceived autonomy but allowed sufficient opportunity for customization and improvisation of the process and/or the outcome.

IMPLICATIONS FOR DESIGNING CREATIVE EXPERIENCES

Taken together, the results from the studies described in this chapter create a significant challenge for those who are trying to design creative experiences for others. Put simply, people are most likely to use their most creative cognitive processes when multiple constraints are active; however, these conditions may not be the most conducive for maximizing self-perceptions of autonomy or competence. Thus, careful thought must be taken when designing a creative experience through which another is intended to navigate in order to provide the most favorable conditions for a positive experience and outcome.

To make this point more vividly, consider the example of consumer co-production, a growing phenomenon in the global economy. Recent technological advances have enabled consumers to collaborate much more closely with companies to create products, services, and experiences. "Co-creation," "co-production," and "self-design" enable consumers, generally through web-based toolkits, to act as the designers of a product to suit their individual preferences. The product is then manufactured exclusively for them. These products are offered in a diverse set of categories, from sportswear (Nike), backpacks (LL Bean), candy (M&Ms), computers (Dell), condiments (Heinz), cell phone covers (Mytego), and lamps (Access Artisans). In most cases, the consumer who chooses to design their own product pays a premium for the chance to do so, and as such, the companies designing the web sites need to understand better how to enhance the consumer's design experience. The research presented here suggests some possible tools that companies might use to help guide consumers take on this new role as designer.

Avoiding the Path of Least Resistance: Managing the Outcome

It may seem like an optimal strategy for a company that allows and encourages consumers to design their own products to provide examples of others that have been designed and professionally produced as well. However, if the individual's primary goal is to create a genuinely unique outcome for themselves, the strong tendency to follow a path of least resistance suggests otherwise. Instead, consumers may be more likely to produce genuinely unique, novel outcomes if they are provided with an blank slate that shows them what decisions they need to make, but does not inadvertently suggest what those decisions should be. Nike iD, for example, does this by giving consumers the chance to design their own shoe. The consumer starts from a rough sketch (black lines on a white background), and fills in the different areas of the shoe with selected colors from a color palate.

Designers of the consumer experience could also ensure a more nearly unique outcome by actively blocking the availability of certain options at different points during the design process. If consumers were heading down a standard path, about to design a product that many others had created before them, the technology could either actively warn the consumer about their intent to design a rather ubiquitous product or it could more passively ensure that the consumer would not make those choices.

Maintaining Perceptions of Autonomy and Competence

In either case, the interaction with the consumer could be designed in such a way that the consumer still felt in control of the design experience and decisions (to maintain autonomy) and felt proficient enough to navigate through the process (to maintain competence). Research has demonstrated that when the design tasks require attribute-related knowledge (e.g., designing a laptop computer), companies should design the creative experience differently for novices than for experts (Randall, Terwiesch, and Ulrich, 2005, 2007). Specifically, when constructing their products, novices should select the different components based on needs, whereas experts should be allowed to choose according to the actual parameters. Our research suggests that similar assessments may be necessary based on the consumers' self-reported design ability or confidence. Some people may prefer more autonomy and less guidance whereas, other may prefer the balance to swing in the other direction. As technology changes dramatically, the experience can be adapted to an individual's preferences. Such a customized experience ensures not only satisfaction with the unique, creative outcome but also with the novel experience as well.

Conclusion

Despite the recent flurry of creativity research in both marketing and psychology, according to Sternberg and Dess (2001), "we do not know enough about this important psychological process" (p. 332). Certainly that statement also applies to our understanding of consumers' information processing and underlying motivations during creative tasks. While restricted in its scope, our research is designed to initiate a more thorough examination of consumer creativity from both a cognitive and a social-psychological perspective. By focusing on the influence of constraints, which are common contextual factors in consumption situations, we were able to examine consumers at the individual level and gain some insights regarding the influence of constraints on two critical aspects of creative tasks: the outcomes produced and the experience itself.

References

ALBA, J. W., and HUTCHINSON, J. W. (1987). Dimensions of consumer expertise. *Journal of Consumer Research, 13* (March), 411–454.

BARON, R. M., and KENNY, D. A. (1986). The moderator-mediator variable distinction in social psychological research: Conceptual, Strategic, and Statistical Considerations, *Journal of Personality and Social Psychology*, 51 (6), 1173–1182.

BETTMAN, J. R., LUCE, M. F., and PAYNE, J. W. (1998). Constructive Consumer Choice Processes. *Journal of Consumer Research, 25* (Dec.), 187–217.

CORBIN, J., and STRAUSS, A. (1990). Grounded theory research: Procedures, canons, and evaluative criteria. *Qualitative Sociology, 13*(1), 3–21.

COSTELLO, F., and KEANE, M. (2000). Efficient creativity: Constraint-guided conceptual Combination. *Cognitive Science, 24*(2), 299–349.

CSIKSZENTMIHALYI, M. (2000). The costs and benefits of consuming. *Journal of Consumer Research, 27* (Sept.), 267–272.

DAHL, D. W., and MOREAU, C. P. (2005). Constraints and creative enjoyment. Unpublished manuscript.

DAHL, D. W., and MOREAU, C. P. (2007). Thinking inside the box: Why consumers enjoy constrained creative experiences. *Journal of Marketing Research, 44*(3), 357–369.

DECI, E. L., and RYAN, R. M. (2000). The "what" and "why" of goal pursuits: Human needs and the self determination of behavior. *Psychological Inquiry, 11*(4), 227–268.

FINKE, R., WARD, T., and SMITH, S. (1992). *Creative cognition: Theory, research, and applications.* Cambridge, Mass.: MIT Press.

Fournier, S., and Mick, D. G. (1999). Rediscovering satisfaction. *Journal of Marketing, 63* (Oct.), 5–23.

Gentner, D. (1989). The mechanisms of analogical transfer. In Vosniadou and Ortony (Eds.), *Similarity and analogical reasoning* (pp. 99–124). Cambridge, U.K.: Cambridge University Press.

Glaser, B. G., and Strauss, A. (1967). *The discovery of grounded theory: Strategies for qualitative research.* Chicago: Aldine Publishing Company.

Guilford, J. P. (1950). Creativity. *American Psychologist, 5,* 444–454.

Hennessey, B. A. (2000). Self-determination theory and the social psychology of creativity. *Psychological Inquiry, 11*(4), 293–298.

Johar, G. V., Holbrook, M. B., and Stern, B. B. (2001). The myth of creative advertising design: Theory, process, and outcome. *Journal of Advertising, 30*(2), 1–25.

Kirmani, A., and Campbell, M. C. (2004). Goal seeker and persuasion sentry: How consumer targets respond to interpersonal marketing persuasion. *Journal of Consumer Research, 31*(4), 573–582.

Moreau, C. P., and Dahl, D. W. (2005). Designing the solution: The impact of constraints on consumer creativity. *Journal of Consumer Research, 32* (June), 13–22.

Murphy, G. L. (1988). Comprehending complex concepts. *Cognitive Science, 12*(4), 529–562.

Newell, A., and Simon, H. A. (1972). *Human problem solving.* Oxford, England: Prentice-Hall.

Perkins, D. N. (1981). *The mind's best work.* Cambridge, Mass.: Harvard University Press.

Randall, T., Terwiesch, C., Ulrich, K. T. (2007). User design of customized products. *Marketing Science, 26*(2), 268–280.

Randall, T., Terwiesch, C., and Ulrich, K. T. (2005). Principles for user design of customized products. *California Management Review, 47*(4), 68–85.

Rossant, J. (1996). Somewhat individual. *Forbes, 157*(2), 152.

Ryan, R. M., and Deci, E. L. (2000). Self-determination and the facilitation of intrinsic motivation, social development, and well-being. *American Psychologist, 55*(1), 68–79.

Sanbonmatsu, D., Kardes, F., and Herr, P. (1992). The role of prior knowledge and missing information in multiattribute evaluation. *Organizational Behavior and Human Decision Processes, 51*(1), 76–92.

Sheldon, K. M., and Elliot, A. J. (1999). Goal striving, need satisfaction, and longitudinal well being: The self concordance model. *Journal of Personality and Social Psychology, 76*(3), 482–497.

Sternberg, R. J. and Dess, N. K. (2001). Creativity for the new millennium. *American Psychologist, 56*(4), 332.

Stokes, P. D. (2001). Variability, constraints, and creativity: Shedding light on Claude Monet. *American Psychologist, 56*(4), 355–359.

Wallendorf, M., and Belk, R. (1989). Assessing trustworthiness in naturalistic consumer research. In Elizabeth Hirschman (Ed.), *Interpretivist Consumer Research* (pp. 69–84). Provo, Utah: Association for Consumer Research.

Ward, T. B. (1994). Structured imagination: The role of category structure in exemplar Generation. *Cognitive Psychology*, *27*, 1–40.

Ward, T. B. (2001). Creative cognition, conceptual combination, and the creative writing of Stephen R. Donaldson. *American Psychologist*, *56*(4), 350–354.

White, R. W. (1959). Motivation reconsidered: The concept of competence. *Psychological Review*, *66*, 297–333.

CHAPTER 7

THE DEVELOPMENT AND EVALUATION OF TOOLS FOR CREATIVITY

STEVEN M. SMITH

ANDRUID KERNE

EUNYEE KOH

JAMI SHAH

Tools and other artifacts can be seen as extensions of our human selves. For example, hand tools can be seen as extensions of the hand's ability to grasp, strike, or dig, and vehicles extend the ability of our legs to take us places. Likewise, information technology (IT) tools can extend and support the limited cognitive systems and abilities of humans. For example, memory storage systems, from writing tablets to books to digital memory devices, vastly extend the limits of our long-term memories. The use of computer windows, or any device with active files, can foreground information far beyond the limitations of our human working memory capacity, functionally extending this important cognitive ability. Cognitive systems upon which we rely every day include lower-order cognitive systems, such as sensation, perception, pattern recognition, working memory, and long-term memory; as well as higher-order cognitive systems, such as language

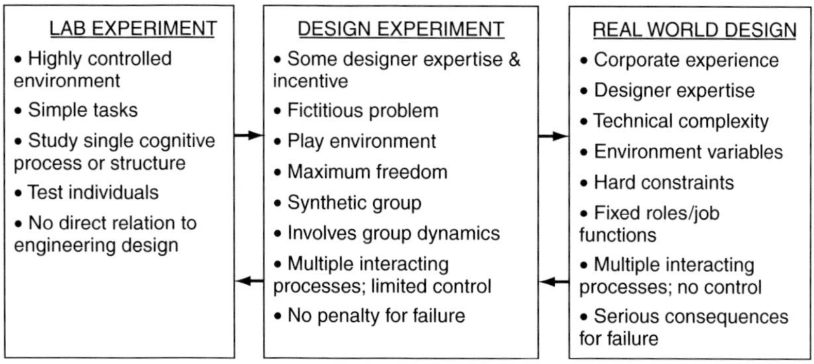

Figure 7–1 Alignment of research across levels of complexity and ecological validity.

systems, concept formation, visualization, and reasoning. What does cognition have to do with creativity? How can IT tools extend and enhance the cognitive systems and abilities that give rise to creative ideas? How can the efficacy of IT tools for enhancing creativity be analyzed and tested empirically? The present paper addresses these questions.

Creativity can be defined as anything made by people that is in some way novel and has potential value or utility. The creative cognition approach to understanding creativity posits that, although creativity depends upon many important factors, such as personality, environment, and historical settings, it also depends critically upon cognition (Finke, Ward, and Smith, 1992; Smith, Ward, and Finke, 1995). First of all, the creative cognition approach states that cognition, itself, has inherently creative qualities (Smith, Ward, and Finke, 1995). For example, language is not simply a reflexive system for parsing and transmitting information; rather, we create new utterances and comprehend novel phrases many times every day. Memory, far from being a passive, recorded repository, involves creative construction and reconstruction, routinely creating new memories to make sense of past episodes. Furthermore, the role of cognition in creative work is critical. Cognition that is commonly seen to be involved in creative work includes set-breaking (which can be enhanced by contextual shifting; e.g., (Smith, 1995), intuitive guiding (Bowers, Farvolden, and Mermigis, 1995), conceptual combination and extension (Ward, 1995), transfer of analogies from remote domains (Gick and Holyoak, 1980), and visual synthesis (Finke, Ward, and Smith, 1992). Although understanding the role of cognition may not be sufficient for understanding creativity, it is nonetheless necessary.

Empirical studies of creative cognition do not necessarily examine creativity as a whole, but rather focus on the cognitive processes and structures that collaborate in the production of creative ideas (Shah et al., 2003). Although it is desirable to know whether or not a technique or a method or a tool enhances creativity, one can often make more progress if one tries to determine exactly if and how various components of a method or tool affect aspects of cognition. For example, rather than simply asking whether or not brainstorming enhances creativity, one might test whether or not an instruction to avoid criticizing ideas leads to the generation of more remotely associated ideas, or whether group idea generation causes more fixation than does individual idea generation. Such an approach can do more than simply give a "thumbs-up" or "thumbs-down" sign for a method or tool; it can guide the development and improvement of methods, techniques, and tools for enhancing creativity.

Fixation and Incubation in Laboratory Studies

In the course of creative thinking, several classic phenomena are often observed. Two such phenomena that are related to each other are fixation and incubation. We have experimentally studied these creative cognitive phenomena in a variety of tasks, ranging from tightly controlled laboratory studies of memory and problem solving, to field studies of engineering designers working on realistic design tasks. Well-controlled experimental studies are needed to clearly establish causal relations among variables, whereas field studies are necessary to furnish the ecological validity of the observed phenomena. Finding that similar effects occur across studies that vary in their levels of complexity and ecological validity is referred to as *alignment* (Smith and Blankenship, 1989), and it is this alignment process that allows us to infer that ecologically valid phenomena occur because of known cognitive mechanisms. We now describe an example of this research alignment process in relation to the phenomena of fixation and incubation.

Incubation in creative problem solving is a mysterious and remarkable phenomenon. The term *incubation* refers to cases in which taking time away from one's work can result in surprising flashes of insight. There are many colorful examples in which incubation effects resulted in historically important discoveries. For example, Archimedes, stumped on approaches to determining the volume of an irregularly shaped crown, had a flash of insight when getting into the bath, resulting in his discovery of the displacement principle. Henri Poincaré was stepping onto a bus on a holiday trip when he experienced a flash of insight leading to his discovery of the Fuchsian Functions. Ludwig van Beethoven, while dozing in his carriage on the way to a concert, experienced a flash of insight for a musical canon. Nobel Prize–winning chemist Kary Mullis was driving through the countryside one evening when a double insight furnished him with the two key ideas behind his invention of the polymerase chain reaction (PCR). NASA scientist Jim Crocker, taking a shower in his hotel room, had a flash of insight for a method for spacewalking astronauts to repair the myopic Hubble Space Telescope.

In spite of these and other well-documented cases of incubation, as well as countless anecdotal instances, incubation proved elusive to experimental studies until recent years. Replicable incubation effects in the laboratory were not found until the phenomenon was linked to an initial period of experimentally induced fixation (Shah et al., 2003).

Blocker	Fragment	Target
ANALOGY	A _ L _ _ G Y	ALLERGY
BRIGADE	B _ G _ A _ E	BAGGAGE
COTTAGE	C _ T A _ _ G	CATALOG
CHARTER	C H A R _ T _	CHARITY
CLUSTER	C _ U _ T R _	COUNTRY
CRUMPET	C U _ P _ _ T	CULPRIT
DENSITY	D _ _ N I T Y	DIGNITY
FIXTURE	F _ I _ U R E	FAILURE
HOLSTER	H _ S T _ R _	HISTORY
TONIGHT	T _ N G _ _ T	TANGENT
TRILOGY	T R _ G _ _ Y	TRAGEDY
VOYAGER	V O _ _ A G E	VOLTAGE

Figure 7–2 Materials used in implicit memory blocking. The fragment for each 7-letter target word had letters in common with the corresponding blocker word, but could not be completed correctly by the blocker. Reprinted with permission from Smith, S.M., & Tindell, D.R. (1997). Memory blocks in word fragment completion caused by involuntary retrieval of orthographically similar primes. *Journal of Experimental Psychology: Learning, Memory and Cognition*, 23(2), 355–370.

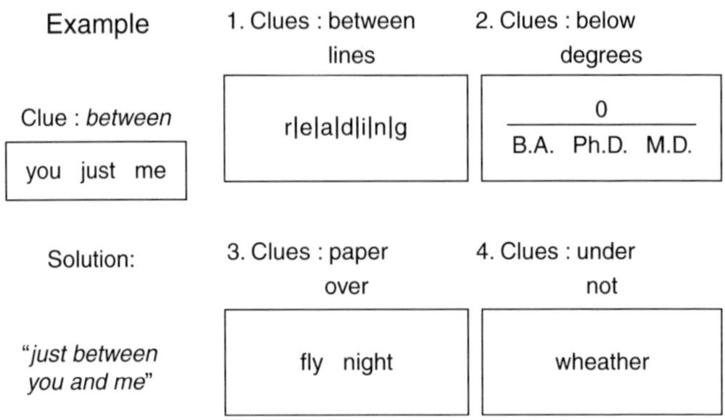

Figure 7–3 The solution to each rebus, or picture-word puzzle, was a common English phrase. Clues for non-critical rebuses (numbers 1 and 2) were helpful, encouraging participants to use the provided clues on the critical problems. The misleading clues for critical rebuses suggested wrong answers to the problems. Reprinted with permission from Smith, S.M., & Blankenship, S.E. (1989). Incubation effects. *Bulletin of the Psychonomic Society*, 27, 311–314.

This view, sometimes referred to as the *forgetting fixation hypothesis*, describes incubation as the culmination of a sequence of cognitive states. Beginning when one takes on a problem to solve, the problem is initially represented in the solver's mind, and initial work on the problem begins. If this initial work brings the problem to completion, or if the problem is ultimately unsolvable by the problem solver, then it is not a candidate for incubation (Smith, 1994; Smith, 1995). In other cases, the initial work can reach an impasse, known as *fixation*. It is at this point of fixation that a break from the problem becomes an instrumental element in the incubation process; after enough time, or with shifts in the problem-solving context, the fixated approach to the problem becomes less dominant. This decrease in fixation allows the problem solver to form an altered representation of the problem, one that omits or bypasses the fixated approach and that can trigger insight into the solution to the problem. Note that this theory differs from theories that postulate that a break from a fixated problem allows autonomous unconscious processes, step by step, to bring work on a problem to its completion.

Remote	Associates	Test problems	Blockers	Solutions
SALAD	HEAD	GOOSE	*lettuce*	egg
BED	DUSTER	WEIGHT	*room*	feather
APPLE	HOUSE	FAMILY	*green*	tree
CAT	SLEEP	BOARD	*black*	walk
WATER	SKATE	CUBE	*sugar*	ice
ARM	COAL	STOP	*rest*	pit

Figure 7–4 Each three-word Remote Associates Test problem had a one-word solution that was associated with each of the three corresponding test words, whereas each blocker was associated with only two of the three test words. Reprinted with permission from Smith, S.M., & Blankenship, S.E. (1991). Incubation and the persistence of fixation in problem solving. *American Journal of Psychology*, 104, 61–87.

Our research on this subject has focused on fixation effects, and to a lesser extent on the effects of breaks from problem solving or from creative work. In pursuit of alignment of research efforts across levels of complexity and ecological validity, we have cast these experiments within a range of tasks, including memory tasks, simple problem solving, playful creative idea-generation tasks, and realistic conceptual-design tasks.

In implicit memory tasks, one's use of prior knowledge and experiences occurs automatically, involuntarily, and without conscious intentions. Even amnesiacs who have little ability to consciously remember recent events nonetheless can show intact implicit memory. For example, a patient with anterograde amnesia might not be able to recall seeing the word ANALOGY after a few minutes, but would nonetheless find it easier to complete the word fragment A _ _ L _ GY after seeing the solution word. People with normally functioning memory show the same implicit memory effects in word-fragment completion. Implicit memory can also block performance on this task, such as when the word ANALOGY is followed a few minutes later by the word fragment A _ L _ _ G Y. Because one's implicit memory

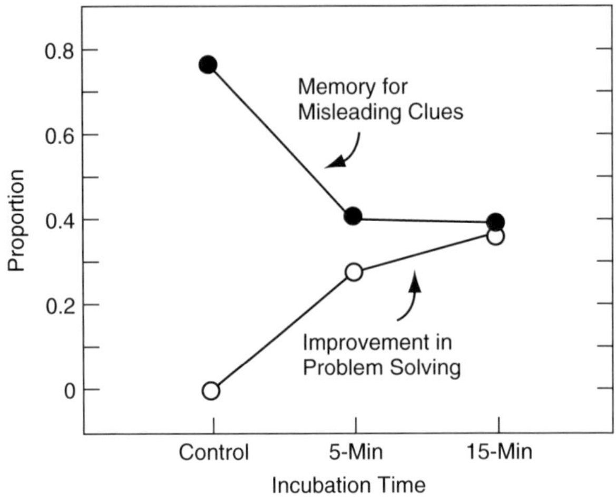

Figure 7–5 The longer that re-testing of initially unsolved critical problems was delayed, the fewer misleading clues were recalled, and the more likely participants were to resolve the initially unsolved problems. Reprinted with permission from Smith, S.M., & Blankenship, S.E. (1989). Incubation effects. *Bulletin of the Psychonomic Society,* 27, 311–314.

automatically provides the recently encountered word ANALOGY, which does not solve the fragment, but nearly fits the solution, an involuntary memory-blocking effect is observed (the correct solution is ALLERGY), as shown by in several experiments reported by Smith and Tindell (1997). Even when participants were warned that previously viewed words could not complete the test word fragments, the implicit memory-blocking effect was not diminished. Other examples of implicit memory-blockers are shown in Figure 7–2.

In simple problem solving the same pattern is observed: stimuli that were recently encountered that seem related to problem solutions are inappropriately brought to mind for rebus problems (Smith and Blankenship, 1989) (see Fig. 7–3) and Remote Associates Test problems (Smith and Blankenship, 1991) (see Fig. 7–4).

Furthermore, as these misleading "clues" are forgotten over time away from initially unsolved problems, people are better able to resolve the fixated problems (Figs. 7–5 and 7–6). This incubation effect, documented repeatedly, shows that incubation is causally linked with initial fixation in problem solving.

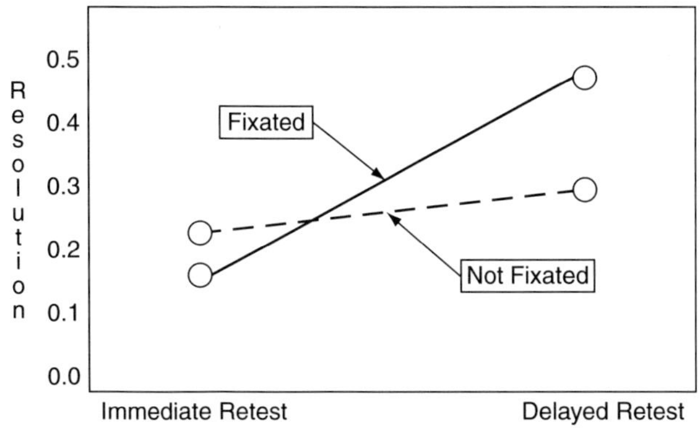

Figure 7–6 Initially unsolved Remote Associates Test problems were re-tested either immediately or after a delay. Re-testing after a delay increased resolution rates, but only for initially fixated problems that had been accompanied by blockers. Reprinted with permission from Smith, S.M., & Blankenship, S.E. (1991). Incubation and the persistence of fixation in problem solving. *American Journal of Psychology*, 104, 61–87.

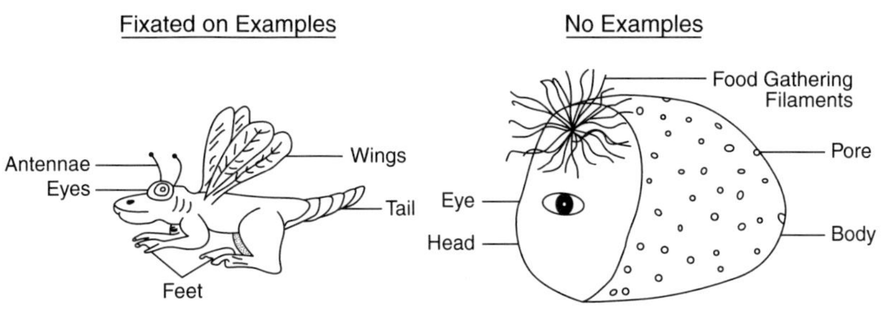

Figure 7–7 Life forms drawn by students in study by Smith et al; the sketch on the left is by a student who saw examples, the sketch on the right by a student who saw no examples. Reprinted with permission from Smith, S.M., Ward, T.B., & Schumacher, J.S. (1993). Constraining effects of examples in a creative generation task. *Memory & Cognition*, 21, 837–845.

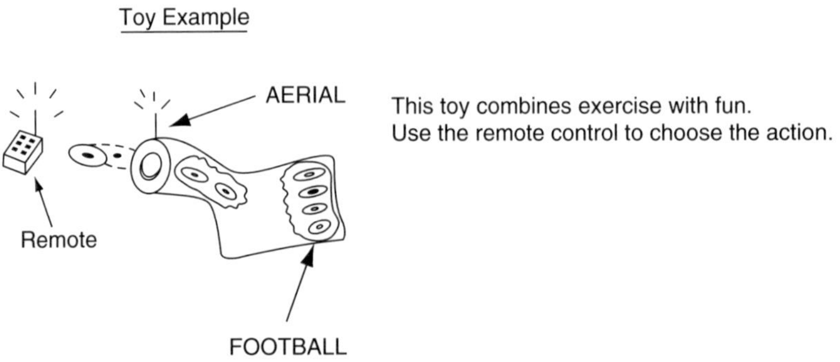

Figure 7–8 Example of Toy from Smith et al. Each of the three examples used electronics, a ball, and a high level of physical activity. Reprinted with permission from Smith, S.M., Ward, T.B., & Schumacher, J.S. (1993). Constraining effects of examples in a creative generation task. *Memory & Cognition*, 21, 837–845.

Experimentally induced fixation effects caused by recent viewing of examples can also be seen in more open-ended tasks that involve creative idea generation, also known as *ideation*. Smith, Ward and Schumacher (2003) devised a creative idea-generation task in which participants were asked to create, sketch, and describe novel ideas that they had not previously encountered. In one version of this ideation task, undergraduate students were asked to spend an hour sketching and labeling life forms that might evolve on a planet similar to Earth (Fig. 7–9). In a second version of the task, students were asked to invent, sketch, and describe new toys that they had never seen or heard of before (Fig. 7–8). In these experiments, before the participants began, they were presented either three similar examples of ideas for novel creatures (or toys), or, in a control condition, they were given no examples. Participants' ideas were scored by counting the number of ideas generated (this frequency metric was never significantly influenced by the manipulated variables), and each generated idea was scored according to the presence or absence of each of the three critical features of the examples. For the life-form–generation task, all three examples had four legs (critical feature #1), antennae (critical feature #2), and a tail (critical feature #3). The critical features in the toy-generation task, found in all three examples, were electronics, a ball, and a high level of physical activity. The results were scored separately for each critical feature by assessing the probability that a participant's ideas included a

Figure 7–9 Example of Life from Smith, Ward, and Schumacher. Each of the three examples had four legs, antennae, and a tail. Reprinted with permission from Smith, S.M., Ward, T.B., & Schumacher, J.S. (1993). Constraining effects of examples in a creative generation task. *Memory & Cognition*, 21, 837–845.

given critical feature, as well as by a measure of overall similarity to examples, which was the mean of the probabilities of all three critical features. These experiments, and many that followed, found that student participants incorporated significantly more of the features of the examples in their creative sketches if they had seen the examples, as compared to a control group who received no examples (see Fig. 7–7). Even when students were instructed to create ideas as different from the examples as possible, fixation on the viewed examples did not diminish. Thus, use of the fixating examples in creative ideation, like implicit memory-blocking, is difficult to escape.

Measuring Creative Ideation in Engineering Design

More expert participants—engineering design students and professional designers—have also been shown experimentally to fixate on problematic examples. After viewing the example of a measuring cup for the blind shown in Figure 7–10, students tended to design devices that were highly similar to the example, and notably, like the example, were non-infinitely variable, and lacked overflow mechanisms (Jansson and Smith, 1991).

Figure 7–10 Example of measuring cup for the blind from Jansson & Smith. The example has problematic features, including the fact that it is non-infinitely variable, and that it lacks a needed overflow device. Reprinted with permission from Jansson, D.G., & Smith, S.M. (1991). Design fixation. *Design Studies*, 12 (1), 3–11.

This *design fixation* was even found when the designers were explicitly instructed to avoid flaws in the examples. Designers in Jansson and Smith's (Jansson and Smith, 1991) study who were shown the example of a cheap, spill-proof coffee cup shown in Figure 7–11 were told not to use straws or mouthpieces in their inventions. Nevertheless, exposure to the flawed example greatly increased the occurrence of these flaws in the designs. Jansson and Smith observed design fixation even in professional engineering designers.

Not only design fixation, but incubation effects have also been observed in engineers. Engineering design students working on a design project for an advanced class worked with their teams either in back-to-back sessions, or with a day's break between sessions. Design ideation metrics derived by Shah, Vargas-Hernandez, and Smith (2003) were used to score the projects. These metrics included measures of quantity (number of ideas generated by a participant), variety (number

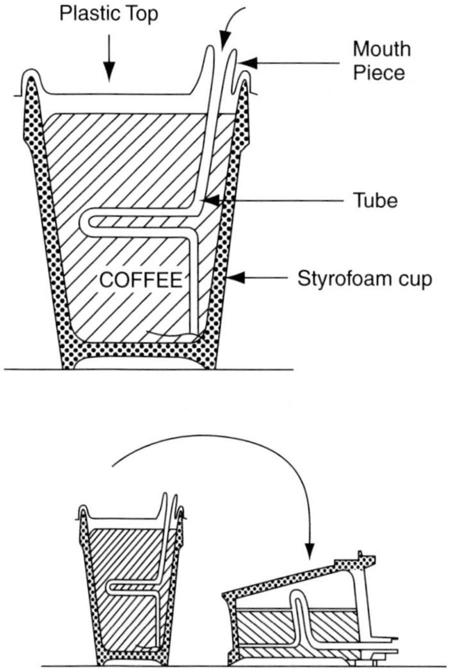

Figure 7–11 Example of spill-proof coffee cup from Jansson & Smith. The example uses a straw and a mouthpiece; instructions forbade the use of these features. Reprinted with permission from Jansson, D.G., & Smith, S.M. (1991). Design fixation. *Design Studies*, 12 (1), 3–11.

of different categories of ideas generated by a participant), quality (this measure was subjectively scored by teaching assistants using clear criteria laid out by the professor for the design class), and novelty (average statistical infrequency of ideas relative to a norm). In order to assess novelty, it was first necessary to compile a norm from all ideas generated by all of the experimental participants, and calculate the frequency with which each idea was generated. Novelty scores for each idea were derived by dividing an idea's normative frequency by the total number of ideas generated for the norm. The mean novelty for each participant's ideas was analyzed as a function of whether or not participants had seen the fixating example. Variety scores were derived from the same norm: after categorizing all ideas according to their respective categories, the number of categories of ideas for each participant was used in the analysis.

Examples of participants' ideas are shown in Figure 7–12. The same metrics were used in a parallel study of divergent thinking, using non-expert students given the task of listing ideas for uses for a two-liter plastic soda bottle. These students worked individually for two 15-minute sessions, with the sessions either back-to-back, or separated by a 20-minute break. Thus, both studies manipulated incubation and used the same ideation metrics, but at very different levels of complexity and ecological validity. The two experiments showed parallel results with respect to incubation: in both experiments, all four measures of ideation improved as a function of incubation, relative to no incubation (Fig. 7–13).

These experimental studies establish the presence of fixation across a broad range of tasks, from highly controlled artificial laboratory tasks to complex ecologically valid tasks involving design and invention. They also show that incubation, a component of methodological tools for supporting creative ideation, has a beneficial effect across these same levels.

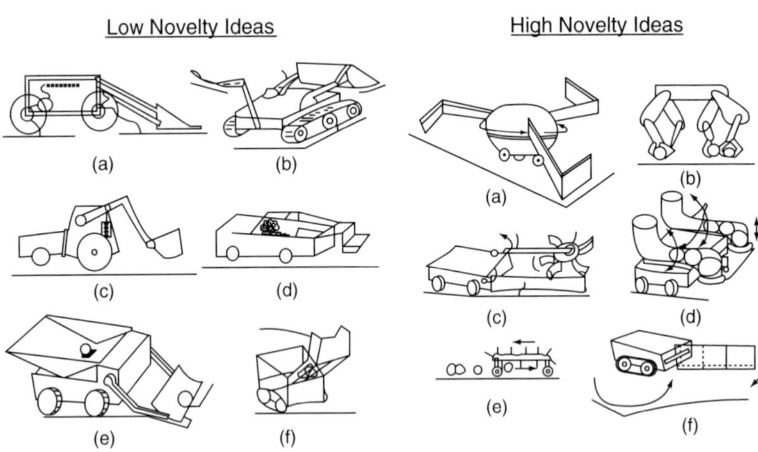

Figure 7–12 Examples of high and low novelty ideas generated by participants in Shah et al. Reprinted with permission from Shah, J.J., Smith, S.M., Vargas-Hernandez, N. (2003). Metrics for measuring ideation effectiveness. *Design Studies*, 24, 2003, 111–134.

Measure	Quantity	Variety	Quality	Novelty
Laboratory Experiment: Divergent Thinking Scores				
Control	1.51	1.20	1.47	2.82
Incubation	1.93	1.60	2.37	3.73
Design Experiment: Ideation Effectiveness Scores				
Control	4.86	2.81	6.15	4.71
Incubation	5.11	6.24	7.31	6.76

Figure 7–13 Ideation and incubation at different levels of complexity & ecological validity. The means are shown for the measures of quantity, variety, quality, and novelty.

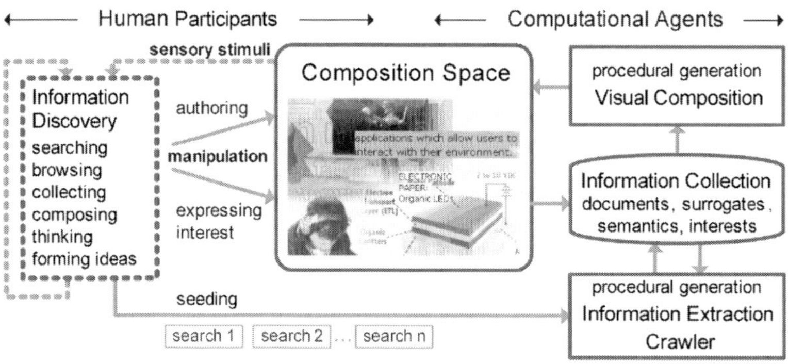

Figure 7–14 Computational agents support human participants engaged in information discovery with combinFormation.

The Information Discovery Framework

We build on the ideation measurement tools to investigate creativity and how it can be supported in human interactions with digital information. The information discovery framework (Kerne et al., 2006) extends creative cognition to enable empirical testing and analysis of the efficacy of IT tools for enhancing creative ideation, the cognitive systems, and abilities that give rise to creative ideas. Whereas others address the role of search technologies in exploration (White, Kules, Drucker, and Schraefel, 2006), information discovery inverts the focus to investigate activities in which people need to develop new ideas and engage in information finding. The representation shifts associated with insight and ideation, such as changes in conceptual framing and information needs, are the crux of information-discovery knowledge-creation tasks, such as invention or the formulation of a thesis topic. Found information stimulates one to see new perspectives and formulate new mental models.

The design ideation metrics of quantity, variety, quality, and novelty previously discussed are used to measure creativity in individual ideas produced by test participants. For information-discovery tasks, we developed a new measure for variety, in addition to categorizing ideas and counting the categories addressed by each participant. To measure the diversity of encountered information, we count the number of information resources to which the participant navigates. To further extend ideation metrics, information discovery developed a metric for another component of ideation, *emergence*, which addresses composite ideas that form from combinations of individual ideas. Emergence is measured by assessing how a participant develops coherent groups of informational and ideational elements, and the insight and novelty within that characterize such groups (Kerne et al., 2008).

Supporting Emergence with combinFormation

We utilize the information discovery framework in studying the effectiveness of *combinFormation*, an IT tool for supporting and enhancing emergence through combinatorial play. Development of the combinFormation creativity support tool was initiated with the intention of bringing the methodological approaches of postmodern artists and composers working in diverse media into everyday interactive experiences with digital information. Instead of creating original masterworks from scratch, artists of this period used "referencing" to assemble works that make extensive use of citation. Marcel Duchamp (Lippard, 1972) and John Cage (1961) developed work by changing the contexts, and thus the meanings, of *found objects*. We take the same approach to information finding. Visual artists such as Max Ernst (Spies, 1998) developed the medium of *collage* as a provocative means of intentionally presenting conceptually oriented collections. Filmmakers since Sergei Eisenstein have juxtaposed clips to create *montage* (Eisenstein, 1942); again, focusing on how meaning changes through contexts of combination. Sound composers from Karlheinz Stockhausen to DJ Spooky (2004) *remix* found sounds, leveraging the recontextualization of combination. Guy Debord (1981) applied these concepts to develop *detournement* in the context of social action. We connect the common methodological threads of these modern and postmodern artists to form the concept of *recombinant information* (Kerne, 2005; Kerne et al., 2008). The essential underpinning is that the juxtaposition and recontextualization of elements leads to new readings, new understanding, and thus to the emergence of new ideas. Found objects, collage, montage, remix, and *detournement* are essential forms of information recombination, to which we now add mixed-initiative information composition.

combinFormation[1] utilizes the form of *composition*, which visually and conceptually integrates elements to represent a collection through information recombination. In *mixed-initiative composition*, people work with software agents to build visual semantic collections (Fig. 7–12). System agents extract clippings from documents, which function as surrogates, and assemble them visually and procedurally (see example, Fig. 7–14). The visual composition is procedurally generated over time, like a dynamic video. Related surrogates are spatially clustered (Kerne, Koh, Sundaram, and Mistrot, 2005). Procedural generation iteratively places visual representations into the composition space, where the participant can see and manipulate relationships among them. This can stimulate cognitive restructuring and creative ideation. Design tools are available in the context of the element, providing capabilities for creating and creating personal collections as navigable compositions. Colors, sizes, fonts, and compositing can be adjusted. Compositing creates visual blends, contrasting with the cut-and-paste adjacency juxtaposition style of hard edges and clear lines, which is better for representing relationships among elements, while maintaining individual characteristics.

The combinFormation seeding interface enables participants to input multiple search queries and select a specific search provider for each, such as Google, Yahoo News Search, Flickr, or del.icio.us. The agent uses the

[1] combinFormation is freely available for use at http://ecologylab.cs.tamu.edu/combinFormation/

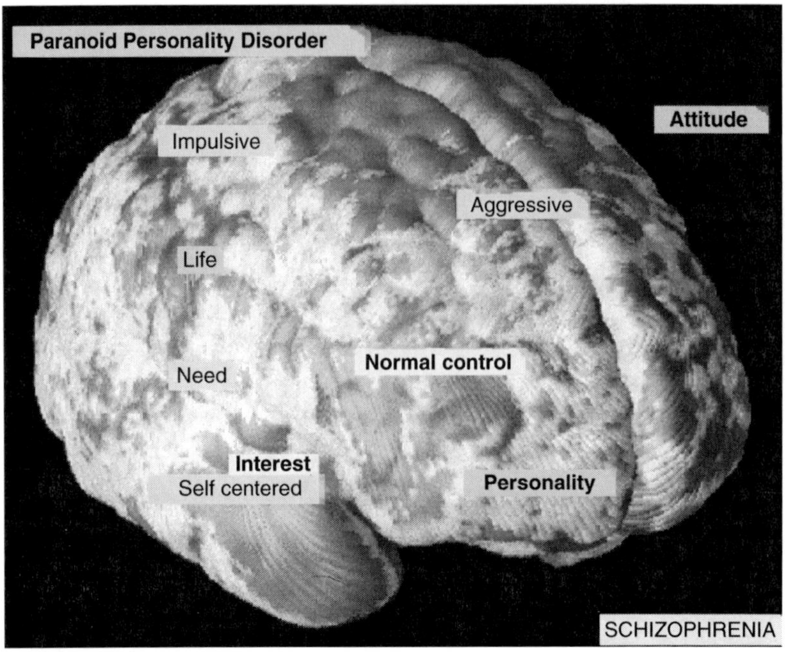

Figure 7–15 Composition of image and text surrogates representation answer to the dating information discovery question. This was scored as emergence 3, quality 2. This answer juxtaposes an interesting set of psychological factors relating to dating with a picture of the brain, creating a sort of map. The juxtaposition is provocative and thoughtful, leading the viewer to think about how parts of the brain might relate to these issues in the dating experience. This is highly emergent. There is not, however, a very clear explanation of the interconnections shown, resulting in a lower score for quality.

seeding specifications as the initial basis for procedural information extraction. It processes search result documents, extracting image and text clippings that function as surrogates. The participant engages in processes of searching, browsing, collecting, and authoring media in the composition space, which serves as a visible medium for communication between human and agent, as well as for thinking about and sharing information resources. When the participant brushes a surrogate in the composition space with the mouse, semantic metadata details-on-demand are visualized in-context. Participants can directly experience the juxtaposed surrogate clippings, and they can also navigate back to source documents for more in-depth information. While browsing and manipulating surrogates, they can use a fluid interface to express interest, directing the human-in-loop system to retrieve more relevant information.

Participants save their collections as compositions as XML, JPG, and HTML files. They reopen what they saved in combinFormation with the XML file to continue exploring and refining. They publish or e-mail their HTML and JPG files to easily share their collections-as-compositions.

Studying the Effects of combinFormation on Information Discovery

We are conducting a series of laboratory and field studies on how working with combinFormation affects information discovery, with significant alignment of results. In all of the studies, we assess components of the creative products of students engaged in information discovery tasks, which require searching, collecting, and conceptually connecting information. We compare how the students use combinFormation in one condition with how they use the Google search interface in conjunction with a text-based tool, such as Microsoft Word.

A laboratory study utilized a reduced version of combinFormation to show that representing collections with composition improves information discovery (Kerne et al., 2007; Kerne et al., 2008). A quantitative field study found that undergraduates performed better on projects in a course on innovation and invention when they collected prior work with combinFormation (Kerne et al., 2006). In a qualitative investigation of the experiences of these students, they reported that combinFormation's representation of information collections with mixed-initiative visual compositions provided provocative stimuli that helped them overcome fixation (Cage, 1961).

Laboratory Study of Emergence

When building complex systems, isolating the impact of components and independently assessing their efficacy is imperative. As creative cognition breaks creativity into a set of components to be independently measured, we conducted a study to isolate and investigate the efficacy of one component of combinFormation: the composition of image and text surrogates representation. Our central hypothesis was that the composition of image and text surrogates representation would increase emergence during the performance of information discovery tasks. In the experimental scenario, undergraduate psychology students answered

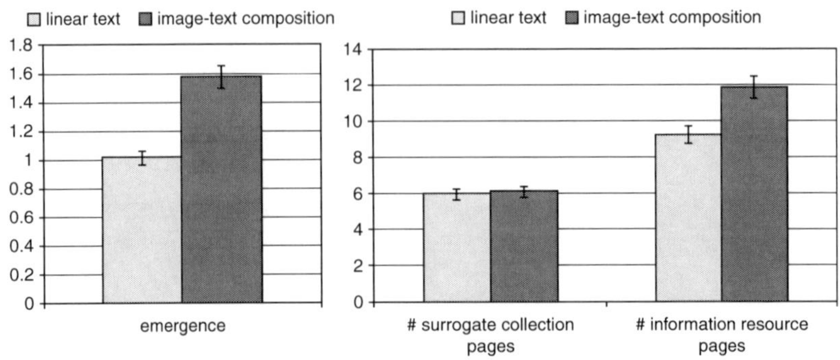

Figure 7–16 Left: The mean emergence measure (scale 0–3), as differentiated by the representation format condition; Right: Navigational Variety and Efficiency: Per participant avg. number of surrogate collection pages and avg. number of information resource pages by representational format. Reprinted with permission from Kerne, A., Smith, S. M., Koh, E., Choi, H., Graeber, R., An Experimental Method for Measuring the Emergence of New Ideas in Information Discovery, *International Journal of Human-Computer Interaction*, 2008, *24*(5), 460–477.

open-ended information discovery questions about life experience, such as, "What psychological factors can influence a person's experiences dating?" To form answers the students were asked to create collections of information surrogates, annotated with their own thoughts.

Our experiment included a limited, direct-manipulation-only version of combinFormation. All proactive agent components were turned off. We curated a source collection of psychology resources. In one experimental condition, the source collection representation was a set of compositions of image and text surrogates that we had previously created using combinFormation. In this condition, the students also used limited combinFormation to create their answers. In the other experimental condition, both the source and answer collection representations utilized typical linear text to author their creative products. Each student answered two information discovery questions with one apparatus, and two with the other.

Information discovery ideation metrics were applied to assess the creative products. Information representation was shown to significantly impact the emergence and variety ideation metrics (Figure 7–16) (Kerne et al., 2007; Kerne et al., 2009). Figure 7–15 shows one student's answer to the dating question.

Quantitative Field Study in the Design Process

We designed and conducted a comparative field study in the Design Process course. Students used either combinFormation or Google and Word to collect prior works for their Hybrid and Invention assignments. Half the class was assigned to use the mixed-initiative composition system, combinFormation, for the prior work collection on The Hybrid, with the other half the class using Google to search and Word to assemble relevant results (Google + Word). For The Invention, the groups switched. Thus, each half of the class used combinFormation (mixed-initiative composition) for one assignment, and Google + Word (textual list) for the other. This was fair to students, while providing comparative conditions of information representation for study.

The course's Teaching Assistant (TA) evaluated both components of the assignment—the prior work and the creative products—for both projects. The criteria and process for evaluating the creative products were established in the Design Process course in prior years, before combinFormation's introduction there. For the creative invention products, the criteria were originality, novelty, practicality, broad impact, and commercial transfer potential. For the prior work, the Design Process

	cF	Google +Word	not involved	total/ average
# of students	59	88	35	182
% of students	32.4%	48.4%	19.2%	100%
prior work	3.08	2.32		2.63
Hybrid	3.32	2.85		3.04

	cF	Google +Word	not involved	total/ average
# of students	61	98	23	182
% of students	33.5%	53.9%	12.6%	100%
prior work	3.13	2.38		2.66
Invention	3.41	2.85		3.06

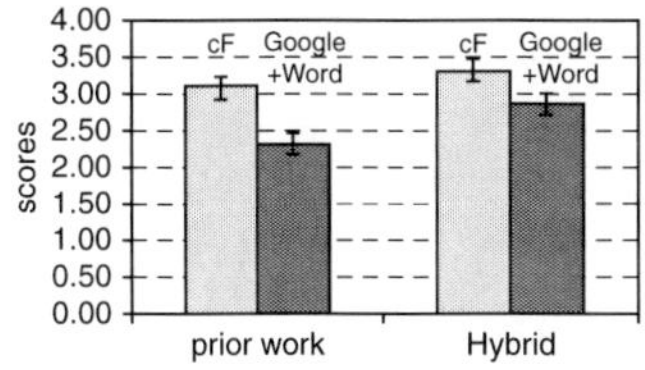

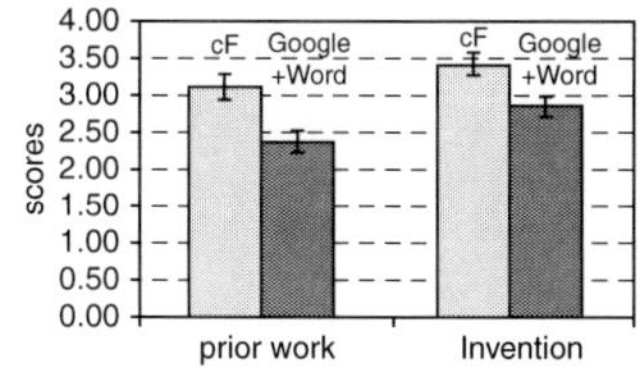

Figure 7–17 Left: Student scores on the Hybrid assignment; Right: Student scores on the Invention assignment. Reprinted with permission from Kerne, A., Koh, E., Dworaczyk, B., Mistrot, M.J., Choi, H., Smith, S.M., Graeber, R, Caruso, D., Webb, A., Hill, R., Albea, J. (2006). combinFormation: A Mixed-Initiative System for Representing Collections as Compositions of Image and Text Surrogates, Proc Joint ACM/IEEE Conference on Digital Libraries (JCDL) 2006, 11–20.

course and combinFormation research teams collaborated to establish criteria for evaluation: how informative, communicative, and expressive, the collection was, and the variety among the collected resources. For both components of both assignments, a new 1–5 scale was used for the study. This scale corresponds directly to the letter grades that were assigned in the course.

Significant differences were observed across the information representation condition for the prior work and for the creative products on both assignments in the field study (see Fig. 7–17) (Kerne et al., 2006). The results demonstrated that combinFormation better supports students engaged in information discovery tasks in collecting and putting together prior works. According to the scores, the TA found that representations of collections assembled in the

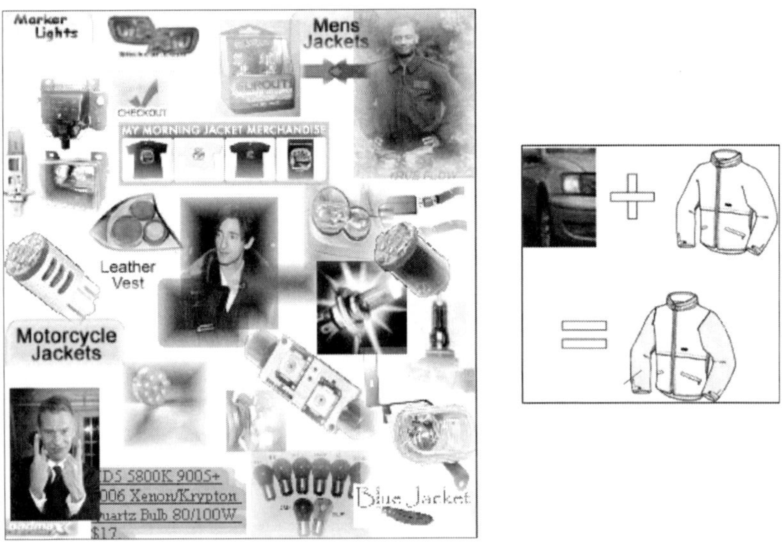

Figure 7–18 Left, prior work collection for collaborative student Hybrid assignment project in The Design Process Course, developed as a composition using combinFormation. Each surrogate is navigable to the source document it was extracted from. In sketch on right, the resulting BlinkerJacket invention addresses bicycle safety by integrating turn signals into clothing. Reprinted with permission from Kerne, A., Koh, E. (2007). Representing Collections as Compositions to Support Distributed Creative Cognition and Situated Creative Learning, *New Review of Hypermedia and Multimedia* (NRHM), 13(2) Dec 2007.

medium of visual composition were better than text lists (Google + Word) for understanding, developing ideas, and communicating meaning. Furthermore, students created better inventions when they used combinFormation to develop prior work. The field study results demonstrated that combinFormation's visual, mixed-initiative method for searching, organizing, and integrating information promotes creative information discovery in education.

Qualitative Field Study in the Design Process

Distributed cognition is a theoretical and methodological framework that constitutes cognitive processes beyond a single brain and body by using the functional relationships of elements that participate in the processes (Cage, 1961). Cognition is embodied and situated through socially organized work activities. Its study has been based primarily on qualitative data. We introduce the term *distributed creative cognition* to address creative ideation processes that occur in distributed environments of participants, artifacts, context, and practice. When cognition is distributed across multiple participants, we need to understand how artifacts and processes contribute to creative ideation. We focus on the role of digital representations, such as the composition space, in promoting the emergence of new ideas.

We studied the best cases in order to understand how using combinFormation for prior work collections contributes to distributed creative cognition in information discovery scenarios of invention and research (Kerne and Koh, 2007). Two exemplary project teams were selected from the Design Process course, based on their Hybrid project scores, and asked, through an email sent only to members, to participate in a group interview about how they developed their invention projects. The members of each group met in person with the interviewer and talked informally, in a semi-structured interview, about their group's invention development process, including use of combinFormation. Figure 7–18 presents a prior work collection from one of the interviewed teams, who developed a project called "Blinker Jacket." Blinker Jacket combines a jacket with turn signals, to address nighttime bicycle safety.

The procedural generation of the composition space by software agents produced effective provocative stimuli. Students said that class projects tended to be similar because many students lived in the same environment on campus. Many invention ideas were based on changing things in this environment, an example of fixation. However, through using combinFormation, students were stimulated to think more broadly about the world, because they saw diverse visual information. The provocative stimuli of the procedural generation of the visual information representations helped them overcome fixation. The composition in Figure 7–18 was the result of seeding combinFormation with the queries "car light" and "blinker jacket."

The role of the human manipulability of information in a composition space relates cognitive restructuring to the interplay between the representation and the act of shaping it. Group members said that when they moved jacket and light pictures around, they could visualize what they could make through combinations. They experienced the generated compositions as messy, but this was a jumping-off point, not a barrier.

Through their embodied interactions with the generated visual information representations, students developed complex relationships with the ideas and each other. The effects of visual representations, procedural generation, and manipulability provoked distributed cognition through dialogue, emergence, and concretization of abstract ideas.

Discussion and Conclusions

Tools for creative innovation should be guided by the need to extend human creative cognition. Tests of tools for creative innovation should analyze effects of multiple components of the tools, and how those components contribute to aspects of creativity. By studying the efficacy of tools like combinFormation in this analytical experimental manner, focusing on enhancements of the cognitive processes that underlie and give rise to creativity, we can develop better tools to support creativity and innovation.

Quantitative methods are available for assessing the products of creative processes, and these can be invoked to evaluate creativity support tools. The invocation of these methods is laborious, involving the development of contextualized protocols for assessing features of the products of particular tasks. These protocols must then be applied first individually and independently, and later collectively and interdependently, to each creative product in each experiment. Unbiased consensus must be developed among experimenters at each stop of the process. Such incremental steps, however, can give us a better footing in terms of knowing what components of a system facilitate creative production, and which do not. Qualitative data add dimension to quantitative results by depicting how components of creative cognition function in practice. Over time, this mixed-method approach (Cage, 1961) will lead to the development of better and more effective tools that reliably support creativity and innovation.

Future development of creativity support tools will investigate other operations of creativity beyond emergence. Other operations that people employ in creative cognition include generating remote associations, transferring analogies, thinking abstractly, visually synthesizing designs, generating divergent ideas, recognizing important clues, and transcending implicit assumptions. Software tools that can support and enhance creative cognition can be designed and developed with these sorts of operations in mind. As with combinFormation, experimental tests of a tool's efficacy should focus on whether the tool enhances performance of the specific cognitive operations for which it was designed, and how the enhancement is achieved.

References

Bowers, K. S., Farvolden, P., and Mermigis, L. (1995). Intuitive antecedents of insight. In Steven M. Smith, Thomas B. Ward, and Ronald A. Finke (Eds.), *The creative cognition approach* (pp. 27–51). Cambridge, Mass.: MIT Press.

Cage, J. (1961). *Silence.* Middletown, Conn.: Wesleyan University Press, 1961.

Debord, G. (1981). *Situationist international anthology* (pp. 55–56). Berkeley, Calif.: Bureau of Public Secrets.

Eisenstein, S. (1942). *The film sense.* New York: Harcourt.

Finke, R., Ward, T., Smith, S. M. (1992). *Creative cognition.* Cambridge, Mass.:, MIT Press.

Frechtling, J., Westat, L. S., Katzenmeyer, C. (1997). National Science Foundation user-friendly handbook for mixed method evaluations, available at http://www.ehr.nsf.gov/EHR/REC/pubs/NSF97–153/START.HTM (accessed on October 1, 2008)/

Gick, M. L., and Holyoak, K. L. (1980). Analogical problem solving. *Cognitive Psychology, 15,* 306–355.

Hollan, J., Hutchins, E., and Kirsh, D. (2000). Distributed cognition: Toward a new foundation for human-computer interaction research. *ACM Transactions on Computer-Human Interaction, 7*(2), 174–196, June.

Interface Ecology Lab. combinFormation, available at http://ecologylab.cs.tamu.edu/combinFormation. (accessed on October 1, 2008)

Jansson, D. G., and Smith, S. M. (1991). Design fixation. *Design Studies, 12*(1), 3–11.

Kerne, A. (2005). Doing interface ecology: The practice of metadisciplinarity. *Proceedings of SIGGRAPH 2005, Art and Animation,* 181–185.

Kerne, A., Koh, E., Dworaczyk, B., Mistrot, M. J., Choi, H., Smith, S. M., Graeber, R, Caruso, D., Webb, A., Hill, R., Albea, J. (2006). combinFormation: A mixed-initiative system for representing collections as compositions of image and text surrogates. *Proceedings of the Joint ACM/IEEE Conference on Digital Libraries (JCDL),* 11–20.

Kerne, A., Mistrot, J. M., Khandelwal, M., Sundaram, V., Koh, E. (2004). Using composition to re-present personal collections of hypersigns. *Proceedings of Computational Semiotics in Games and New Media (CoSIGN).*

KERNE, A., KOH, E. (2007). Representing collections as compositions to support distributed creative cognition and situated creative learning. *New Review of Hypermedia and Multimedia, 13*(2), Dec.

KERNE, A., KOH, E., SMITH, S. M., CHOI, H., GRAEBER, R., WEBB, A. (2007). Promoting emergence in information discovery by representing collections with composition. *Proceedings of ACM Creativity and Cognition.*

KERNE, A., KOH, E., SUNDARAM, V., MISTROT, J. M. (2005). Generative semantic clustering in spatial hypertext. *Proceedings of ACM Document Engineering,* 84–93.

KERNE, A., SMITH, S. M. (2004). The information discovery framework. *Proceedings of ACM Designing Interactive Systems.*

KERNE, A., SMITH, S. M., KOH, E., CHOI, H., GRAEBER, R. (2008). An experimental method for measuring the emergence of new ideas in information discovery. *International Journal of Human-Computer Interaction, 24*(5), 460–477.

KERNE, A., SUNDARAM, V. (2003). A recombinant information space. *Proceedings of Computational Semiotics in Games and New Media (CoSIGN),* 48–57.

LIPPARD, L. (1972). *Dadas on art.* Englewood Cliffs, N.J.: Prentice Hall.

SHAH, J. J., SMITH, S. M., VARGAS-HERNANDEZ, N., GERKENS, R., and WULAN, M. (2003). Empirical studies of design ideation: Alignment of design experiments with laboratory experiments. *Proceedings of the American Society of Mechanical Engineering.*

SMITH, S. M., and BLANKENSHIP, S. E. (1989). Incubation effects. *Bulletin of the Psychonomic Society, 27,* 311–314.

SMITH, S. M., and BLANKENSHIP, S. E. (1991). Incubation and the persistence of fixation in problem solving. *American Journal of Psychology, 104,* 61–87.

SMITH, S. M., WARD, T. B., and SCHUMACHER, J. S. (1993). Constraining effects of examples in a creative generation task. *Memory and Cognition, 21,* 837–845.

SMITH, S. M. (1994). Getting into and out of mental ruts: A theory of fixation, incubation, and insight. In R. Sternberg and J. Davidson (Eds.), *The nature of insight* (pp. 121–149). Cambridge, Mass.: MIT Press.

SMITH, S. M., WARD, T. B., and FINKE, R. A. (1995). *The creative cognition approach.* Cambridge, Mass.: MIT Press.

SMITH, S. M. (1995). Fixation, incubation, and insight in memory, problem solving, and creativity. In S. M. Smith, T. B. Ward, and R. A. Finke (Eds.), *The creative cognition approach* (pp. 135–155). Cambridge, Mass.: MIT Press.

SMITH, S. M., and TINDELL, D. R. (1997). Memory blocks in word fragment completion caused by involuntary retrieval of orthographically similar primes. *Journal of Experimental Psychology: Learning, Memory and Cognition, 23*(2), 355–370.

SHAH, J. J., SMITH, S. M., VARGAS-HERNANDEZ, N. (2003). Metrics for measuring ideation effectiveness. *Design Studies, 24,* 111–134.

SPIES, W. (1998). *Max Ernst collages: The invention of the surrealist universe.* New York: Harry N. Abrams, 1988.

SPOOKY, DJ, (2004). *Rhythm science.* Cambridge, Mass.: MIT Press, 2004.

WARD, T. B. (1995). What's old about new ideas? In S. M. Smith, T. B. Ward, and R. A. Finke (Eds.), *The creative cognition approach.* Cambridge, Mass.: MIT Press.

WHITE, R. W., KULES, B., DRUCKER, S. M., SCHRAEFEL, M. C. (2006). Supporting exploratory search. *Communications of the ACM, 49*(4), 36–39.

CHAPTER 8

CONCEPTNETS FOR FLEXIBLE ACCESS TO KNOWLEDGE

THOMAS B. WARD

EXISTING knowledge is crucial to innovation, yet knowledge can be a double-edged sword. It can provide guidance in formulating and exploring new ideas, but it can also constrain thinking and limit the possibilities individuals and groups consider. One dominant view of how to deal with constraints, as expressed in popularized versions of creativity enhancement, is to throw out everything you know because it must somehow be wrong, and it will block you from seeing creative solutions. My research takes a somewhat different approach and suggests that your knowledge is good; most of what you know is in fact correct and quite applicable to future problems. Being creative and developing innovations depends, not on rejecting what you know, but at least in part, on flexibly accessing and using knowledge from multiple levels of abstraction (Ward, 1994, 1995; Ward, Patterson, Sifonis, Dodds, and Saunders, 2002; Ward, Patterson, and Sifonis, 2004).

In considering the dual role of knowledge in innovation, it is useful to note that creative accomplishments can take many forms. They can range from new conceptualizations that radically alter how we see ourselves and our place in the universe (e.g., the shift from a geocentric model to a heliocentric model), to relatively small tweaks of existing products (e.g., adding a new soft drink flavor to an existing line of carbonated beverages).

Much of what is considered in the current chapter is more toward the latter end of that spectrum, with a particular focus on *conceptual expansion*, the deliberate extending of the boundaries of conceptual domains by mentally generating new domain instances and bringing them to fruition in the form of tangible products. The term *product* is used broadly here to include designs, drawings, descriptions, and prototypes as well as actual, complete, working objects.

Conceptual expansion is a pervasive human activity and can be seen in the crafting of a new monster for a horror novel, the development of a variant on a disposable razor, the design of an experiment to test a psychological theory, the concocting of a new recipe to make leftovers more tolerable, and an innumerable set of other domain extensions. What the activities have in common is that there exists a relatively well-defined domain that contains a set of instances known to the thinker, and the thinker expands the domain by generating a new instance. Conceptual expansion can be thought of as a metal hop (Ward, 1998) rather than a mental leap (Holyoak and Thagard, 1995), but it may nevertheless represent a type of analogical thinking (e.g., Gentner, Holyoak, and Thagard, 2001), in the sense that properties from a well-known entity are projected in a systematic and predictable way onto the new entity. That is, analogy is a process whereby structured knowledge from a well-known source domain in the form of objects, simple relations, and higher-order relations is mapped to a less–well-known target domain in service of understanding, explaining, communicating about, or making inferences about the latter (e.g., positing that, like a solar system, atoms are composed of less massive entities orbiting around a more massive one based on some causal attractive force). In conceptual expansion, the source is often an existing product whose configuration of attributes is projected in some modified or transformed way onto the empty space that will become the new product (e.g., devising a new board game based on the locations, possible paths, movement determiners, pitfalls, and so on, of some specific previous board game).

Although the focus here is on the incremental advances that are characteristic of conceptual expansion, it is important not to confuse the mental distance traveled with the magnitude of the impact of the new idea, product, or process. It should be noted that even modest advances can have enormous economic and societal consequences. For instance, although Edison's lightbulb was a close variant on preexisting designs of which Edison was cognizant (Friedel and Israel, 1986), its success in yielding a reliable, long-lasting source of electric light nevertheless drastically altered our day-to-day lives. More generally, the economy of scale means that tiny incremental advances can have major financial implications. When one sells billions of gallons of soft drinks, for example, an advance that produces a savings of a mere fraction of a cent on each can sold results in a huge increase in profit.

Specific Instances and Abstract Knowledge in Conceptual Expansion

An organizing framework for considering the importance of flexible access to knowledge at different levels of abstraction is the *path-of-least-resistance model* (Ward, 1994, 1995; Ward, Dodds, Saunders, and Sifonis, 2000). The model states that, when people develop new ideas for a particular domain, the predominant tendency is to access fairly specific, basic-level exemplars from that domain as starting points, and to project many of the stored properties of the instances onto the novel ideas being developed. For example, in devising a new sport, the predicted predominant tendency would be for people to retrieve specific known instances of sports, such as baseball and football, and to pattern the new sport after those instances. Following the path is expected to result in reduced originality of the new ideas, in contrast to other, more abstract approaches to accessing knowledge. On the other hand, there may be benefits to relying on specific instances in terms the practicality or feasibility of the new ideas.

There are interesting anecdotal and historical accounts that reveal the possible constraints imposed by relying on specific known instances, as well as the possible advantages. For example, in the 1830s, when passenger rail travel was just getting started in the United States, designers seem to have patterned the first railway passenger cars directly on horse-drawn stagecoaches of the day, including the fact that conductors had to sit on the outside of the car (White, 1978). This approach was efficient in the sense that railway passenger cars became available quickly, but because the conductors were seated on the outside, several of them fell off and were killed. Thus, a property of an existing domain instance that was unnecessary and potentially harmful was nevertheless carried over to the new idea being developed. Another example is that, according to Barker (1993), Sony initially abandoned the development of music CDs because the development team started with long-playing record albums of the day as their model, and projected the property of a 12-inch diameter onto the envisioned new device. The device would have held a large and economically unviable quantity of music. Clearly Sony caught up and overtook many competitors, but the initial reliance on a highly specific known instance of a musical storage device slowed their early progress. As a final example of constraints, anyone who has had to scroll down and then back up in reading a pdf version of a journal article in a format that mimics the two-column arrangement of its hard-copy counterpart knows that copying that exact format into electronic form is less than optimal from the point of view of the reader.

On the other hand, there is ample evidence from historical accounts that many non-problematic advances in a wide range of domains also have been based on a slow, incremental process of patterning new ideas after very specific earlier ones (see, e.g., Basalla, 1988). Edison's exploitation of prior lightbulb designs mentioned earlier is an example. Rather than originating the idea of running electricity through a filament in a closed glass container, Edison's accomplishment was to refine the idea by finding the right materials. Another example noted by Basalla is the close connection between Eli Whitney's cotton gin, designed to separate the seeds from the cotton fiber, and a previously existing device, the charka. The charka worked well with the type of cotton that grew in India, the place of its origin, but not with the type of cotton whose seed was more tightly bound to its fiber, which grew well in the southern United States. Rather than pulling the cotton through two rollers as with the charka, Whitney's device used a roller to pull it through a comb-like device. Either Edison or Whitney could have dreamed up entirely new ways of approaching their problems, but they were successful in exploiting ideas that had come before. Thus, even if it may limit novelty, the approach of generating new products by patterning them closely after existing ones may favor practicality over extreme, but impractical originality.

Laboratory research findings mirror these types of real-world phenomena. First, there is the general finding that, when given the task of devising a new domain instance, people develop products that bear a striking similarity to known domain instances. For example, when asked to envision animals on other planets, the vast majority of college students produce descriptions and drawings that resemble typical Earth animals, including such pervasive properties as eyes, legs, and bilateral symmetry (Ward, 1994), and they do so even when given instructions that encourage more originality (Ward and Sifonis, 1997). In addition, just as the innovators noted previously seem to have been influenced by examples they were exposed to (e.g., stagecoaches, lightbulb designs, the charka), so too are individuals in laboratory studies found to copy features of examples they are exposed to (Marsh, Landau, and Hicks, 1996; Marsh, Ward, and Landau, 1999; Sifonis, Ward, Gentner, and Houska, 1998; Smith, Ward, and Schumacher, 1993), and they do so even when features of the examples are identified as being problematic (Jansson and Smith, 1991). It appears that innovation can be constrained by chronically accessible domain instances as well as those made more accessible through recent exposure.

Laboratory findings also reveal some of the properties of existing conceptual structures that are most influential in guiding the form of new ideas. For the three distinct conceptual domains of animals, fruit, and tools Ward et al. (2002) had separate groups of college students perform a non-creative task of listing all of the domain instances they could think of, and a creative task of imagining, drawing, and describing novel instances of those categories that might exist on another planet. Data from the listing task were used to derive a measure of representativeness, namely "Output Dominance," or the number of participants who listed any given exemplar. Exemplars listed by more people can be taken as more representative of the domain. In the creative imagination task, after producing their novel products, participants described the kinds of things they used as the basis for their ideas, and references to specific domain exemplars (e.g., dogs, hammers, oranges) were tabulated to derive a measure of "Imagination Frequency" for each exemplar. The more people who reported relying on a particular exemplar in the creative task, the higher the Imagination Frequency, and the more that exemplar could be seen as influencing creative generation.

Three important results emerged. For all three domains, more than 60 percent of the participants reported relying on at least one basic-level domain instance as a source of ideas for their creations, and there were strong positive correlations between Output Dominance and Imagination Frequency. That is, there was a strong tendency to rely on basic-level domain instances, and those instances tended to be more the more accessible ones that come to mind most readily. In addition, for all three domains, the participants who reported a reliance on specific domain instances developed products that were rated as less original than those developed by participants who reported other approaches, presumably because the properties of individual exemplars (e.g., dogs) are more specific and constraining than the properties associated with higher levels of abstraction (e.g., two eyes symmetrically placed in the head versus some sort of organs for sensing some type of potentially relevant information).

Additional research supporting the value of avoiding readily accessible instances and accessing more-abstract levels of representation reveals that people can be induced to adopt more abstract approaches in conceptual expansion tasks, and that they develop more-original creations as a result (Ward et al., 2004). For example, participants who were asked to imagine life on other planets developed designs that were more original when they were asked to consider abstract attributes of living things (e.g., the need for nutrition to support biological processes) than when they were asked to keep in mind specific Earth animals or were given no special instructions (Ward et al., 2004). Similarly, procedures that preclude reliance on the most readily accessible specific solutions by imposing constraints have been shown to increase originality (Moreau and Dahl, 2005).

Abstraction is a process of accessing knowledge and representing situations in more general terms to overcome the limitations imposed by specific known instances. To use a pragmatic, real-world example, consider the following hypothetical case. If 20 years ago one defined a business venture as "a store that rents VHS tapes" it might have succeeded briefly, but characterizing the venture that specifically would have led to failure with the emergence of new types of storage devices (e.g., DVDs) and new mechanisms of product delivery, such as surface mail (e.g., Netflix) and the Web (e.g., vongo.com). Nor would it be optimal to represent the endeavor as specifically providing electronic access, as evidenced by Blockbuster gaining an advantage by noting that it has actual physical locations where one can go and not have to wait for a new movie to arrive in the mail. Representing the venture more abstractly as "providing temporary access to stored entertainment media" would presumably allow greater adaptability in the face of such changes.

Although accessing abstract information, in contrast to relying on specific domain instances, is linked to greater originality, it is essential to consider another important ingredient of innovative ideas; namely, their usefulness or practicality in meeting the need at hand. A recent study suggests that reliance on specific instances may be more beneficial in terms of practicality. In particular, when participants were given the task of devising new sports, those who reported relying on specific known sports developed ideas that were rated as more playable than those developed by individuals who reported other, more abstract approaches (Ward, 2008). More generally, originality and playability were significantly negatively correlated. To create a scenario to illustrate why that might be true, consider for example, that "ball" might be part of the representation of the specific sport of "basketball," whereas "object contended for" might be the comparable abstract feature in the higher-level concept, "sport." A new sport patterned on the former might include the very practical object of a ball, whereas one patterned on the latter might include an original, but less sensible object that teams contend for, with the result that it would be judged less playable. In either case, an attribute from the accessed concept is projected onto the new situation, but one fosters practicality whereas the other fosters originality.

Thus, even though individuals who naturally adopt more abstract approaches to creative generation tasks, or who are encouraged via experimental manipulations to do so (e.g., Ward et al., 2004) produce more original outcomes, that originality may come at a cost to practicality (Ward, 2008). It should be noted that originality and practicality (in the sense of appeal to consumers) are not always negatively correlated (e.g., Dahl and Moreau, 2002). Nevertheless, both properties need to be considered in assessing the relative merits of reliance on specific instances versus more abstract levels of knowledge.

In the Ward (2008) study participants also rated their own knowledge about sports and took a brief test of sport knowledge. Sport knowledge was found to be significantly positively correlated with the rated playability of the sports they developed. That is, the more-knowledgeable individuals appear to have been better able to exploit their knowledge in service of devising ideas for sports that others might actually like to play. This suggests that some of the cost of moving away from specific examples might be mitigated if tools for managing different levels of knowledge were made available.

Far from rejecting existing knowledge, idea generation in service of innovation requires its judicious use. Whether emphasizing more specific or more abstract knowledge will be most helpful may depend on the relative value assigned to originality or practicality in the project being undertaken, but it is likely that in most cases accessing multiple levels of abstraction will be helpful. In the next section, I sketch some properties of a tool that might aid in that access.

Electronic Concept-Representation Systems as Aids to Innovation

From the preceding discussion, it appears that a significant challenge in fostering innovation is knowledge management. On one hand, it is clear that the originality of newly developed ideas is greater when people access their knowledge at more abstract levels, and it is also clear that they often do not do so spontaneously; instead retrieving and relying on concrete, specific concept exemplars. Moreover, people can be encouraged to access their knowledge at more abstract levels, with a consequent boost in originality. Therefore, a tool that would facilitate people's access to abstract information would be a valuable addition to the innovator's toolbox. Such a tool should provide a way to prod or jog people's memories, allowing them to consider the more abstract concepts of which the concrete entities they retrieve are a part, and that they might not otherwise think about.

On the other hand, there is some evidence that sticking with more-concrete levels of representation can foster the practicality of the new products that are developed. Consequently, a tool should also allow movement from abstract representations to concrete ones that differ from the person's original starting point, but that nevertheless are meaningfully linked to the goals of the creative generation task.

In addition to jogging people's memories of more general or more specific items of knowledge, a tool should provide a way for people to recognize or discover connections across categories. For example, a person's knowledge might contain the idea that people "contend for objects" in activities other than sports (e.g., business, foraging, relationships), and a tool that prodded a branching off from the domain of interest to other domains by way of such shared relations might allow for greater innovation (as in a sport from Ward [2008], in which competitors climbed trees to collect the largest number of coconuts).

So, an ideal tool would facilitate people's access to their knowledge at multiple levels of abstraction, and preserve, highlight, and suggest connections within and across the levels. It should also be noted that such a tool would be broadly useful, not just for the sense of conceptual expansion considered in this chapter, but for the general issue of how people become stuck in, and may be helped to escape from, ruts in their thinking, as was considered, for example, by Smith and his collaborators (this volume; see also Smith, 1995a, b).

Moreover, the innovation tool contemplated here would also allow people to go beyond their own knowledge. That is, a tool would optimally contain the collective knowledge of a large group of individuals so that, in addition to prompting people to recover connections that exist in their own minds but are not readily accessible, it would also allow them discover connections that are new to them. For example, a person might not know that "curling" is a sport, but might be able to discover that fact by traversing a large, stored knowledge base, and then be able to use the properties of that less typical instance in devising their own.

In short, the innovation tool (or set of tools) I have in mind is one that would facilitate access to conceptual knowledge at multiple levels of abstraction to help people probe their own knowledge; see organizing principles and structures of that knowledge that they might not have considered, or only achieved with great cognitive effort; and discover alternate conceptualizations. A key ingredient would be a large-scale, structured electronic concept-representation system similar to or adapted from the WordNet-like databases developed by George Miller and collaborators (WordNet, n.d.) and their ever-expanding variants. WordNet is a lexical database that contains nearly 150,000 unique noun, verb, adjective, and adverb strings organized into *synsets*, clusters of synonymous or interchangeable words.

WordNet-like systems contain several types of links among words, but an especially useful feature is hierarchical connections to superordinates (*hypernyms*) and subordinates (*hyponyms*). Such connections could be used to facilitate or to prompt *abstraction*, the procedure of accessing knowledge at relatively general levels, which has been shown to result in greater originality of products in creative generation tasks (e.g., Ward et al., 2004). In addition, by guiding exploration of *conceptually related* entries, the system would presumably help tie any new ideas to existing ones in meaningful ways that could increase their practical value, or at least guard against losses in practicality observed when people move away from specific instances (Ward, 2008).

WordNet and Creativity

Veale (2004) has provided support for the idea that WordNet can be used in service of creative functioning. Part of Veale's argument relies on the existence of instances of "function-transforming" polysemy within WordNet. To use one of his examples, *knife* is a hyponym of *edge tool* but also of *weapon*. Those broad classes contain objects with similar properties (e.g., sharp edges) and behaviors (e.g., cutting), but different functions (slicing versus stabbing). Because *knife* is represented as a hyponym of both, it allows us to entertain the possibility that other edge tools might also be used as weapons. That is, by using the already-represented polysemy of *knife*, we can examine other hyponyms of *edge tool*, such as *scalpel*, to consider their utility as weapons, and we can also consider the generalization that edge tools are a subset of the broader category, *weapons*. Recent worries about terrorist activities have certainly led authorities concerned with airport passenger screening to consider just such possibilities and, at least temporarily, ban seemingly innocuous items such as nail scissors from carry-on baggage.

Veale referred to using polysemy and category links in the way described in the previous paragraph as *category broadening*. One can conceptually broaden or expand the domain of *weapon*, not just by inventing new instances, but by reinterpreting already existing objects so that they reasonably fit within the domain. Because, in this particular case, *weapon* is a broader category than *edge tool*, the converse—that other weapons may be used as edge tools—is not necessarily supported. Although envisioning what sorts of objects might serve as weapons is not the noblest of creative pursuits, this example nevertheless illustrates how a system such as WordNet might provide support for conceptualizing objects differently.

Veale (2004) also describes a second procedure for supporting creativity, referred to as *category hopping*. Again, to use Veale's example, one could come to think of using a coffee can as a musical instrument by traversing the following path: *Coffee can* is a hyponym of *tin can*, and *tin can* and *steel drum* are related in the sense that they are both hyponyms of *container*. *Steel drum* is also linked to *tympan-membranaphone-drum*, a hyponym of *percussion instrument*. From a psychological standpoint, using a Wordnet system and manually traversing links might allow the user to realize that, in their own internal knowledge base, they have represented the fact that coffee cans are a specific type of tin can, and that tin cans are, in turn, instances of a still-broader grouping of containers that has a multitude of members, including one sense of "steel drum." Because there are multiple senses of *steel drum*, one of which is linked, ultimately, to the broad grouping of things that are musical instruments, it allows consideration of whether the chain of links justifies thinking of a coffee can something that could be used as a musical instrument. That is, by exploring and exploiting the connections that are directly represented in WordNet, one can discover other, potentially useful links.

Thus, polysemy along with hyponym and hypernym relations allows categories to be broadened and objects conceptualized in different ways, and can provide at least some support for creative functioning. In this regard, it should be noted that generating alternate uses for objects is an ingredient of a number of tests of creative potential (e.g.,

Wallach and Kogan, 1965), so the examples provided by Veale do relate directly to aspects of cognition assumed to be associated with creative behavior.

Useful Properties for a ConceptNet-type Tool

The proposed type of tool would facilitate access to more-abstract levels of knowledge and would also be able to suggest novel specific instantiations of that knowledge. Because the application would be focused on the use of conceptual information rather than on language processing *per se*, it might better be described as a "ConceptNet" rather than a "WordNet." That is, the suggestion here is not meant as an endorsement of WordNet as *the* specific electronic concept representation system to be used, but rather as an example of the type of system that would be valuable. Listed below are initial considerations of some properties that would be useful in such a net.

Free Search and Prompting of Abstractions and Specific Instantiations

A ConceptNet would contain the hierarchical and other linkages represented in WordNet, but it may also add new ones such as typicality of instances within superordinate concepts (more on this property later). The system would allow at least two distinct modes of operation: (1) User-determined free search of the net, in which the user would freely explore any and all links in an unconstrained manner, and (2) Prompting, in which the system would make suggestions of alternate interpretations that are at higher levels of abstraction, that have lower typicality within the superordinate, that are less directly connected to the main concept (more intermediate links) or that have other properties that might be deemed important to increasing the originality of the user's thought about the topic.

Considering abstract ideas is not enough for innovation, however. Those abstractions must be realized in the form of more-specific concrete ideas and products. Because WordNets also have links to subordinates, the system would allow a search for or prompting of specific instantiations of a general principle. For example, having allowed a budding entrepreneur to think of VHS tapes as instances of a more abstract concept such as "storage device," the net could allow movement back down to alternative specific types of storage that might in turn suggest other modes of delivery. There is, of course, no guarantee that a preestablished net would contain the needed links for any given insight or discovery, but the process of flexibly moving from lower to higher levels of abstraction and back again could potentially produce such a connection or even help the innovator realize such a connection in his or her own knowledge that was not yet stored in the net. They could then add that connection to the net (more on this later, too).

Domain Specialization

WordNets are largely based on general knowledge associated with concepts in a particular language group (e.g., native speakers of English). As noted, such nets have been shown to allow some types of creative functioning, such as divergent production (e.g., alternate uses for a shoe; see, e.g., Veale, 2004). However, more highly specialized domain-specific versions are possible (e.g., Bentivogli, Bocco, and Pianta, 2004), and indeed necessary to capture detailed knowledge and appropriate abstractions, thus allowing creative progress within particular domains of innovation. Consider the engineering task of designing a braking system for a new vehicle as an example. Thinking of disc brakes as a specific device for stopping a car might not suggest any particularly innovative new ideas, but thinking of braking a vehicle more abstractly as transforming its kinetic energy makes possible clever new concepts such as recovering some of that energy to recharge the battery when braking, as in current hybrid cars. In the standard version of WordNet, *brakes* are types of *restraining devices*, which in turn are types of *devices*. Although potentially useful, that type of abstraction may not be as helpful in design engineering as one that leads ultimately to the physical principles instantiated in devices (e.g., transforming kinetic energy). Thus domain-specific versions are essential.

Developing domain-specific nets may be accomplished in a manner comparable to developing expert systems and could be based on structured interviews with domain experts. Conceivably, a start could be made on characterizing the individual concepts in a domain and the structure of their interconnections by a more automated search and coding of indexes or other, similar procedures, but in the end, the domain systems would probably be socially constructed by the individuals and collaborative groups who work in the domain.

Integration Across Languages and Domains to Facilitate Collaboration

Innovation is not just an individual phenomenon. High-level tasks require collaboration among many individuals with different backgrounds, possibly operating in different locations and having different first languages. Consequently, nets that merge different conceptual representations would be especially helpful.

WordNets have been developed in a wide range of languages, and there are also examples of integrating across multiple languages, as in the case of EuroWordNet developed by Piek Vossen and collaborators (EuroWordNet, 1999), which includes distinct wordnets in each of seven languages that are interconnected through an index based on the original English-language WordNet developed by Miller and colleagues. Presumably such a system could be used to identify translation equivalents, but also the comparable or diverging larger structures of which they are a part, or subtle differences in the sense of the terms in each group. Thus, an important task for the future is developing integrations across representations. Such integrations could facilitate collaborative problem-solving efforts by individuals from different language and cultural groups. That is, by traversing links, not just within one's first language, but also links connecting words in that language to their equivalents (or partial equivalents) in a second language, individuals can broaden their perspectives and develop a keener sense of similarities and differences across the languages. In addition, by jointly traversing and discussing such cross-language links, collaborators with different native languages could be aided in avoiding miscommunications that could hinder group productivity.

Broadening the notion of integration across similar, but non-identical nets, much progress in technology and science involves work at the margins of disciplines and requires collaborative input from specialists who may be unfamiliar with specific meanings of terms in one another's domains, or who may even use similar or identical terms in crucially different ways. Similarly, product development teams may consist of a variety of specialists in marketing, manufacturing, packaging, and so on. In effect, even in the same country and culture with the same native language, collaborative groups may be composed of individuals who work in different specialty areas, each of which has its own special jargon. By analogy to the cross-language case noted in the previous paragraph, jointly traversing and discussing cross-specialty links contained in integrated ConceptNets could lead members of collaborative groups to better understand one another's terminology and perspectives, to work together more cohesively, to avoid miscommunication, and possibly even to gain new insights (by noticing subtle commonalities and differences).

Thus, as noted in the previous section, building separate representations for each specialized domain is an important first step in using ConceptNets for innovation, but a hugely labor-intensive effort to develop nets that integrate terminology and concepts across domains is a critical next step to support interdisciplinary collaboration.

Dynamic Properties and Recording of Paths

Once constructed for individual domains and integrated across collaborative domains, the nets could serve as guides for individuals new to an area; but they should also be completely dynamic in the sense that the connections envisioned and stored by one collaborative group might not meet the specific needs of another. Subsequent groups should be able to modify links and possibly feed those modifications back into the more general system, either as variants that represent alternate paths or as new default paths. Predetermined connections would greatly facilitate entry into a boundary field and allow new contributors to add to the mix earlier than would be possible otherwise, but again, these connections would be seen as guides, not rigid constraints.

So, one important feature of a net would be the ability to track and store the paths that people follow as they work on a given task, and to mark those stored paths in various ways. On the simplest level of marking, all paths taken through the network could be marked as ones that have been explored in the past. The storing of path information could be done dynamically during a given problem-solving session so that individuals could be alerted if they begin to traverse a path they had already taken in the session, but the path information could also be preserved permanently as a record of the ways in which the network has been explored. This would allow individuals who subsequently use the system to choose those same paths or to shun them in favor of novel ones. In addition, permanently stored paths could be marked to indicate the quality of the outcome they provided on previous problems, whether leading to productive outcomes or to blocked thinking. Marking the paths taken and the outcomes obtained would allow future innovators to exploit productive paths and avoid dead-end paths, or to choose entirely new ones in an effort to find completely new directions or perspectives on a problem.

Representing Additional Types of Information

Although WordNet is, to date, a verbal construction, there is no reason in principle why it could not be extended to include visual, spatial, or other types of information. Such extension could be useful in a single system, but it could also facilitate integration across language, culture, or expertise groups. Just as a simple example adapted from Paradis (1979): the English word "*ball*" refers to small items, such as tennis balls, as well as larger ones, such as basketballs, whereas the rough French equivalent "*balles*" is more restricted to the smaller types that fit in one hand. Although such difference could be noted in words, it could also be conveyed by a series of pictures of the kinds of things that would be assigned the respective labels. Search would still be be based on words, but stored pictures linked to the words would provide additional visual information to clarify concepts.

Even within a language group, pictures might help represent subtle variations in the way terms are thought about. In the some regions of the United States, "*tree*" might be represented by a picture of an oak or maple, whereas in West Texas, the most salient representative might be the diminutive mesquite. To the extent that such variations can be captured, cooperative problem solving in groups composed of individuals with diverging backgrounds can be facilitated. Although words could be used to capture the differences, there may be some things that are better represented visually. In addition, diagrams might be used to capture certain interactions among concepts, and movement could also be represented. That is, a net, being virtual and electronic, need not be confused with a static, hardbound textbook or encyclopedic reference source.

Representing Graded Structure Properties

Being socially constructed and with room for individual variation to capture different people's realities, ConceptNets might also contain probabilistic information instead of purely deterministic or absolute information. Much as some dictionaries contain usage information as preferred by x percent of a panel, so could the connections in a ConceptNet contain such information (e.g., "x percent of informants in this group characterize A as an instance of B, but a smaller percentage (y) also think that A could be characterized as a C").

Weighted links based on association norms, output dominance or typicality of category exemplars, centrality of attributes, direction of causality, or any other types of data that characterize people's conceptual systems, could also be represented. Including this information would allow a search for either more or less typical, more or less representative, etc., items in a domain.

Having said that typicality of hyponyms within their hypernym should be incorporated into a WordNet-like system raises the difficult question of how such a feat would be accomplished. Although typicality has been a traditional focus in the categorization literature (e.g., Rosch and Mervis, 1975) and continues to be of much interest (Hampton and Cannon, 2004; Murphy and Ross, 2005), empirical data from participant ratings cover only a tiny fraction of the concepts represented within WordNet. Thus, some more automated means of computing typicality, perhaps from Google or Wikipedia searches, would be needed.

There have been efforts to extend the utility of WordNet entries by computing additional properties, such as the semantic distances between word pairs (Maki, McKinley, and Thompson, 2004). Maki et al. produced a corpus of almost 50,000 word pairs and their semantic relatedness, using, in part, edge information contained in the WordNet database. It is not clear what comparable information, if any, could be used in computing typicality within a hypernym.

SUMMARY

The suggestion for an innovation tool is a knowledge database (or rather a set of such databases) with hierarchical as well as other types of links among concepts that would allow people to flexibly access and use knowledge at multiple levels of abstraction. The systems would allow free search and also prompt users with suggestions. Nets can be general-knowledge versions that might facilitate creativity in domains such as fiction writing, as well as domain-specific ones for innovation in particular specialties. Nets should capture knowledge structures in a particular language or domain, but can also be integrated across languages and domains to facilitate collaborations among groups of individuals from divergent backgrounds. Nets should be highly structured and serve as guides, but also be dynamic and allow storage of new connections and marking of particular types of paths through the database. They should also go beyond words and absolute links to represent visual materials, diagrams, and movement, as well as relative links based on typicality and other aspects of graded conceptual structure.

References

Barker, J. (1993). *Paradigms: The business of discovering the future.* New York: HarperBusiness.

Basala, G. (1988). *The evolution of technology.* London: Cambridge University Press.

Bentivogli, L., Bocco, A., and Pianta, E. (2004). ArchiWordNet: Integrating WordNet with domain-specific knowledge. In P. Sojka, K. Pala, P. Smrz, C. Fellbaum, and P. Vossen (Eds.), *Proceedings of the 2nd Global WordNet Conference* (pp. 39–46). Brno, Czech Republic, January 20–23.

Dahl, D., and Moreau, P. (2002). The influence and value of analogical thinking during new product ideation. *Journal of Marketing Research, 39*, 47–60.

EuroWordNet (1999). Retrieved November 21, 2006, from http://www.illc.uva.nl/EuroWordNet/#EuroWordnet.

Friedel, R., and Israel, P. (1986). *Edison's electric light: Biography of an invention.* New Brunswick, N.J.: Rutgers University Press.

Gentner, D., Holyoak, K., and Kokinov, B. (2001). *The analogical mind: Perspectives from cognitive science.* Cambridge, Mass.: MIT Press.

Holyoak, K., and Thagard, P. (1995). *Mental leaps: Analogy in creative thought.* Cambridge, Mass.: MIT Press.

Maki, W. S., McKinley, L. N., and Thompson, A. G. (2004). Semantic distance norms computed from an electronic dictionary (WordNet). *Behavior Research Methods, Instruments, and Computers, 36*, 421–431.

Marsh, R. L., Landau, J. D., and Hicks, J. L. (1996). How examples may (and may not) constrain creativity. *Memory and Cognition, 24*, 669–680.

Marsh, R. L., Ward, T. B., and Landau, J. D. (1999). The inadvertent use of prior knowledge in a generative cognitive task. *Memory and Cognition, 27*, 94–105.

Moreau, C. P., and Dahl, D. W. (2005). Designing the solution: The impact of constraints on consumers' creativity. *Journal of Consumer Research, 32*(1), 13–22.

Paradis, M. (1979). Language and thought in bilinguals. In W. McCormack and H. Izzo (Eds.), *The Sixth LACUS Forum* (pp. 420–431). Columbia, S.C.: Hornbeam Press.

Sifonis, C. M., Ward, T. B., Gentner, D., and Houska, M. (1997). Relation versus object mapping in creative generation. *Proceedings of the Nineteenth Annual Conference of the Cognitive Science Society, 19*, 1050.

Smith, S. M. (1995a). Fixation, incubation, and insight in memory and creative thinking. In S. M. Smith, T. B. Ward, and R. A. Finke (Eds.), *The creative cognition approach.* Cambridge, Mass.: MIT Press.

Smith, S. M. (1995b). Getting into and out of mental ruts. In R. J. Sternberg and J. Davidson (Eds.), *The nature of insight.* Cambridge, Mass.: MIT Press.

Smith, S. M., Ward, T. B., and Schumacher, J. S. (1993). Constraining effects of examples in a creative generation task. *Memory and Cognition, 21*, 837–845.

Veale, T. (2004). Paths to creativity in lexical ontology. In P. Sojka, K. Pala, P. Smrz, C. Fellbaum, and P. Vossen (Eds.), *Proceedings of the 2nd Global WordNet Conference* (pp. 220–225). Brno, Czech Republic, January 20–23, 2004.

Wallach, M. A., and Kogan, N. (1965). *Modes of thinking in young children.* New York: Holt, Rinehart, and Winston.

Ward, T. B. (1994). Structured imagination: The role of category structure in exemplar generation. *Cognitive Psychology, 27,* 1–40.

Ward, T. B. (1995). What's old about new ideas? In S. M. Smith, T. B. Ward, and R. A. Finke (Eds.), *The creative cognition approach*. Cambridge, Mass.: MIT Press.

Ward, T. B. (1998). Analogical distance and purpose in creative thought: Mental leaps versus mental hops. In K. Holyoak, D. Gentner, and B. Kokinov (Eds.), *Advances in analogy research: Integration of theory and data from the cognitive, computational, and neural sciences*. Sofia, Bulgaria: New Bulgarian University.

Ward, T. B. (2008). The role of domain knowledge in creative generation. *Learning and Individual Differences, 18,* 363–366.

Ward, T. B., Dodds, R. A., Saunders, K. N., and Sifonis, C. M. (2000). Attribute centrality and imaginative thought. *Memory and Cognition, 28,* 1387–1397.

Ward, T. B., Patterson, M. J., Sifonis, C. M., Dodds, R. A., and Saunders, K. N. (2002). The role of graded category structure in imaginative thought. *Memory and Cognition, 30,* 199–216.

Ward, T. B., Patterson, M. J., and Sifonis, C. (2004). The role of specificity and abstraction in creative idea generation. *Creativity Research Journal, 16,* 1–9.

Ward, T. B., and Sifonis, C. M. (1997). Task demands and generative thinking: What changes and what remains the same? *Journal of Creative Behavior, 31,* 245–259.

White, J. H. (1978). *The American railroad passenger car*. Baltimore, Md.: Johns Hopkins University Press.

WordNet: A lexical database for the English language (n.d.) Retrieved March 19, 2009 from http://wordnet.princeton.edu/.

CHAPTER 9

INNOVATION THROUGH tRaNsFoRmAtIoNaL DESIGN

VIKRAMJIT SINGH

BRANDON WALTHER

KRISTIN L. WOOD

DAN JENSEN

THE history of mankind is a testament of how we encounter and solve problems. No matter how mundane a task, at one extreme, or complicated a task, at the other, we constantly *innovate* to solve problems. Innovation is a process that leads to improvements in technology, methods, and our human existence. In engineering, innovation entails the use of tools and processes that enhance the benefits of existing sciences and technologies. These enhancements, in turn, lead to benefits to societal and individual needs. Without innovation, we lose our identity, our ability to adapt, and our motivation to cause change. Innovation is a constant course of action that allows for the expression of creativity, personality, and discovery.

Engineering methods and tools are used to solve real-world problems, whether we are exploring the endless reaches of space or inflating a bike tire. This chapter describes novel tools in engineering design to enhance and empower creativity, and cause the ideation process to move forward. The goal of this chapter is primarily to introduce a developing methodology for design engineers to use in the advancement of mechanical transforming devices. This methodology provides a "snapshot" of how innovation processes can be improved through the use of analogical reasoning and the use of design principles; i.e., meta-analogies. The development of a systematic and methodological approach for identifying transformations in a device is based on a relational view of system-usage scenarios, respective customer needs, and system-level solutions relating to the needs. This area of transformational design is rich with possibilities to create systems that have neither been contemplated nor even dreamed of in the human experience.

This chapter first introduces the topic of transformers and evaluates the motivation for this research. A brief description of the research approach is included, followed by a description of transformational principles and facilitators that are a driving force for this methodology. The chapter then moves step-by-step through the current iteration of the method in detail and concludes with a novel application of transformation applied to everyday systems. In a local context, this chapter seeks to develop a theoretical basis by which transformer design may be wielded by practicing designers. In the larger landscape, however, this paper illustrates a principled approach for ideation with directed methods. This approach is intended to provide a meta-analogy framework by which designers explore solutions that overcome psychological inertia and provide solution paths that are outside the designer's set of experiences.

What Is a Transformer?

When one is asked about transforming products, what comes to mind may be the mid-80s artistic view of a humanoid robot changing into a land vehicle, air vehicle, or dinosaur. Some of these visions of robots were made popular by the television series "*Transformers*" and their toy counterparts. This concept of a transformer, while potentially limiting, does provide a first-order correlation—an icon, and exemplifies some essential rules of transformers. Based on our research into transforming systems, we define a transformer as *a system that exhibits a state change in order to facilitate a new functionality or enhance an existing functionality* (Singh, Skiles, Krager, et al., 2006), (Skiles et al., 2006). A "state" of a system, for the physical or mechanical domain, is defined as *a specific physical configuration in which a system performs a function*. For example a ladder-chair might have two states: one being its chair configuration and the second state being the ladder configuration.

Transforming products have a much broader functional repertoire than traditional single state products. For example, there are transforming 6-in-1 screwdrivers that can change their functionality by changing their head and bit configuration (for example, from a Phillips-head to a slotted head). Transformer applications are present in a broad range of product domains, from household appliances to applications in unmanned aerial vehicles (UAVs). The Switchblade UAV, currently under development by Northrop Grumman, is designed for long-range and long-duration flight. The aircraft could loiter near enemy territory for over 12 hours, then transform to quickly fly to a target when commanded to do so. The claim is that this reconfiguration redistributes shock waves that accumulate in front of a plane at post-Mach speeds and induce drag. At subsonic speeds, the Switchblade's wings swivel back so that they are perpendicular to the fuselage, much like a conventional plane's. Work is also being done on transforming or morphing wings (Popular Science, DATE TK), (Abdulrahim, Garcia, and Lind, 2005), (Singh, Warren, Putnam, et al., 2006); where the wings undergo transformation to provide added functionality to an airframe, such as change in flight characteristics, gust-resistant operation, increase in flight time, etc.

The advantages of transformers include: added functionality, use of fewer resources (e.g., building materials and fuel), and savings in volume and time; however, transformers may also have disadvantages, such as more initial time to develop and complexity in their design (Singh, Skiles, Krager, et al., 2006). It is the role of a transformer design theory to identify when and if transformers should be conceptualized for a given problem, accentuate the advantages, and minimize or remove the disadvantages.

MOTIVATION

With a context of the potential and impact of transforming devices, we focus on where transformation can be used or proven beneficial. To advance the design process, a basic, consistent method is needed to assist in identifying and targeting potential areas for transformation within the design space of a product and in its realm of use. There are, of course, a number of current decision-making design tools. However, none of them focus on a strategy to identify potential domains for state change (or transformation) as a solution to needs or requirements of the design (Pugh, 1996), (Otto and Wood, 2001). There are various design tools that can be used in different phases of a design process, like problem-context questions for gathering customer needs (Otto and Wood, 2001), (Green, 2005), quality function deployment (QFD) for identifying important engineering parameters, theory of inventive problem solving (TIPS) (Sushkov, Mars, and Wognum, 1995), (Altshuller, 1984) to help conceptualize innovative ideas for design conflicts, function structures (Pahl and Beitz, 1999), design structure matrices (DSM) (Boothroyd, Dewhurst, and Knight, 2002), and modular function deployment (MFD) (Skiles, 2006) to identify modules in the design of a product and organize product development tasks or teams. These design tools are not specifically suited to address the design, especially ideation, of transformers directly. For example, they do not explicitly identify different states that could accomplish different functions, nor do they even attempt to identify how the system might transform between these different states. These types of questions are the focus of our present work.

The possible design space in the realm of transformers is just beginning to be examined and appreciated. There are pervasive examples including fixed-wing planes that can fly and hover, or structural beams that extend or collapse to new geometries for different purposes. When we consider these examples, there are prominent questions that come to mind: What are the key needs driving the development of such transformers? Why do designers use a singular state for some products or systems, and when should additional states and transformation be considered? The work presented in this chapter focuses on understanding and answering these questions by beginning to formulate a systematic methodology for designing such systems. The method outlined in this chapter is a work in progress where further research (in the realm of functionality, for example) continues to be applied in order to advance the transformational design theory.

Research Approach

The research approach for this project followed a unique combination of an inductive approach and subsequent deductive reasoning to validate the theory (Singh, Skiles, Krager, et al., 2006). This combined approach, at a high level, is shown in Figure 9–1. The inductive approach is a bottom-up approach where existing transforming systems, in nature (biological

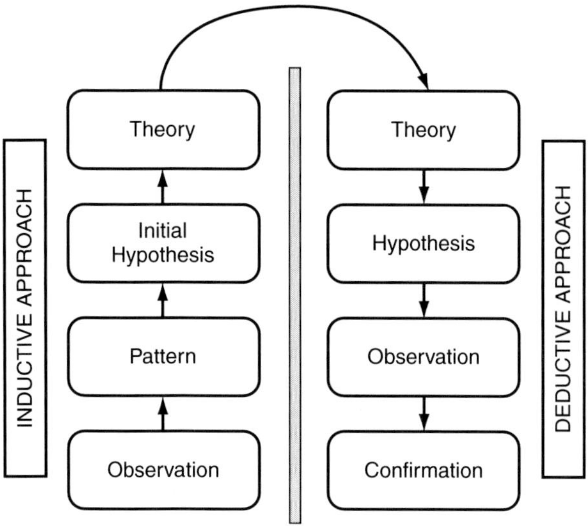

Figure 9–1 Research approach – Inductive with Deductive.

systems), patents, and products, were studied to derive governing transformation design heuristics, referred to as "transformation principles and facilitators." The inductive approach amounts to an empirical study, where the collected data are current or historical transforming systems that exist in nature or were human-generated through serendipitous or ad hoc approaches. The deductive approach, which is a top-to-bottom approach, was simultaneously applied to postulate principles or fundamental concepts, and we subsequently categorized the combined set of validated principles from both approaches. This alternative approach proved to be valuable in creating a method for analyzing product requirements and identifying transformation.

This combined research approach is used to derive heuristic rules or "principles" for transformation from repeated examples found in nature, existing products, and patents that exhibit transformation (inductive approach) and from situations or scenarios that would require the need for transforming a device (deductive approach). Using the combined inductive/deductive approach, we developed a more detailed research study process flowchart that is divided into two sections, where one section follows the inductive approach and the other section the deductive approach. This research flow is shown in Figure 9–2.

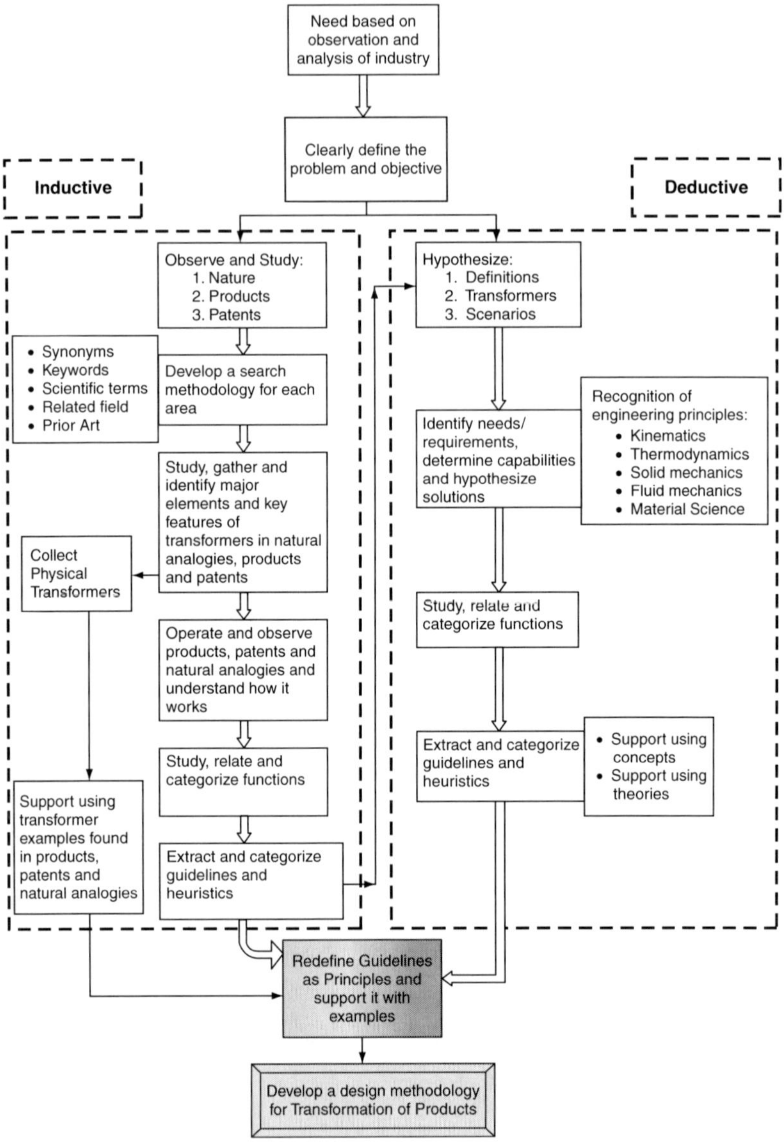

Figure 9–2 Detailed research study process figure.

Transformation Principles and Facilitators

As a result of our initial research (Singh, Skiles, Krager, et al., 2006), (Skiles et al., 2006), (Singh, Warren, Putnam, et al., 2006), (Ericsson and Erixon, 1999), we created a set of governing transformation heuristics. These heuristics help generate physical transformation in a design. These design heuristics are categorized and defined as *transformation principles* and *transformation facilitators.*

Transformation Principle

A *transformation principle* is a generalized directive to bring about a certain type of mechanical transformation. In this sense, it is a guideline that, when embodied, singly creates a transformation. Many embodiments are possible from a given principle, leading to the concept of transformation principles as "meta-analogies."

Transformation Facilitators

A *transformation facilitator* is a design architect that helps or aids in creating mechanical transformation. Transformation Facilitators aid in the design for transformation, but their implementation does not create transformation singly.

Through our research approach as described above, the three (and only three) fundamental transformation principles, which represent transformation potential in the mechanical domain, are: *expand/collapse, expose/cover,* and *fuse/divide.* Subordinate to these three principles are the transformation facilitators. The hierarchical relationship between principle and facilitator exists because principles describe what causes transformation, while facilitators describe what makes the transformation function efficiently and more fully. This category is established through the deductive research process involved in our approach.

The three transformation principles are described below.

- *Transformation Principle #1: Expand/Collapse—Change physical dimensions of an object to bring about an increase or decrease in occupied volume primarily along an axis, in a plane, or in three dimensions.* Collapsible or deployable structures are capable of automatically varying their shape from a compact, packaged configuration to an expanded, operational configuration. For example, portable sports chairs that are now popular expand for sitting and collapse for portability. Puffer fish expand their bodies to ward off and escape predators.

- *Transformation Principle #2: Expose/Cover—Expose a new surface or cover an exposed surface to alter functionality.* This principle is a directive for changing the surface of a device or its parts so as to alter the primary function of the device. This alteration can be brought about by different types of part-to-part interaction of a device and/or the form of the device itself. For example, many cell phones have keyboards that slide out to reveal an operational surface. Similarly, the day-blooming water lily opens during the day to expose its interior.
- *Transformation Principle #3: Fuse/Divide—Make a single-functional device become two or more devices, at least one of which has its own distinct functionality defined by the state of the transformer, or vice versa.* A functional device divides into two or more parts where at least one of the parts has a distinct primary function. Two or more parts with distinct or similar primary functions can fuse/join to form a new device with a different primary function. For example, there are music players that also function as USB flash drives or memory sticks. In nature, army ants join their bodies to form a bridge fro the rest of the colony.

While singly embodying a Transformation Principle can create a transforming product, Transformation Facilitators aid in the design of transformers, but their implementation alone does not create transformation. An example of a Transformation Facilitator is *Common Core Structure*:

- *Common Core Structure—Compose devices with a core structure that remains the same, while the periphery reconfigures to alter the function of the device.* In essence, a reconfigurable device can consist of a core structure that is the main support structure that allows for aligning/positioning different peripheral parts or systems. For example, many leaf blowers can transform to vacuum cleaners by changing the extensions. Reproductive termites begin life as crawling insects, and then grow wings to leave the colony.

Pilot Results of Transformational Principles and Facilitators

The transformation principles and facilitators aid in the design for mechanical transformation. These guidelines, when embodied, help solve design problems by creating a certain type of transformation, thereby acting as a new tool for designers. Using this new tool, a number of transformer concepts were generated and are listed below. Two states for a potential system or product design are shown as the end points between a two-sided arrow. These ideas were generated with a blank canvas, with the transformational principles and facilitators acting as the categories for generating concepts; i.e., mental cues for analogical reasoning.

- Water rocket ← → Squirt gun
- Raincoat ← → Umbrella
- Hose sprayer ← → Lawn/garden sprinkler
- Fishing rod handle ← → Rod stand
- Toaster ← → Electric griddle ← →Cooking top
- Water-sensitive roof shingles ← → Gutters
- Skis ← → Snowboard
- Hairdryer ← → Curling iron ← → Hair straightener
- Headphones ← → Speaker
- Cooler ← → Picnic table

These principles and facilitators were also used as design guidelines in a graduate-level mechanical engineering design course at the University of Texas. The students used mind-mapping (Otto and Wood, 2001) with transformation principles and facilitators to generate innovative concepts. The participants in this exercise were given a prescribed amount of time and a brief tutorial on transformation, the transformational principles, and the transformational facilitators. The students then created a mind map, where the transformation principles become the highest-level categories of the map. This is a slight deviation from the traditional mind-mapping process, in that this technique gives a designer added direction with the inclusion of the transformation principles. As shown in Figure 9–3, each student began the concept generation exercise by writing the three principles. From there, the student generated ideas of potential transforming products that incorporate each transformation principle. The student then wrote the product ideas down by branching the product from the respective principle that aided in generating that particular concept (see Fig. 9–3). Examples of transformer products, as shown in Fig. 9–3, include business shoes that transform to spiked golf shoes, treaded tires that transform to studded tires for snow and ice conditions, and a tote-sized cooler that transforms into a full-sized cooler.

With these innovative and unique results, the potential of the transformational principles and facilitators is illustrated and indicated. The principles and facilitators may serve as invaluable tools to generate concepts that harness the potential of transformation in the mechanical domain.

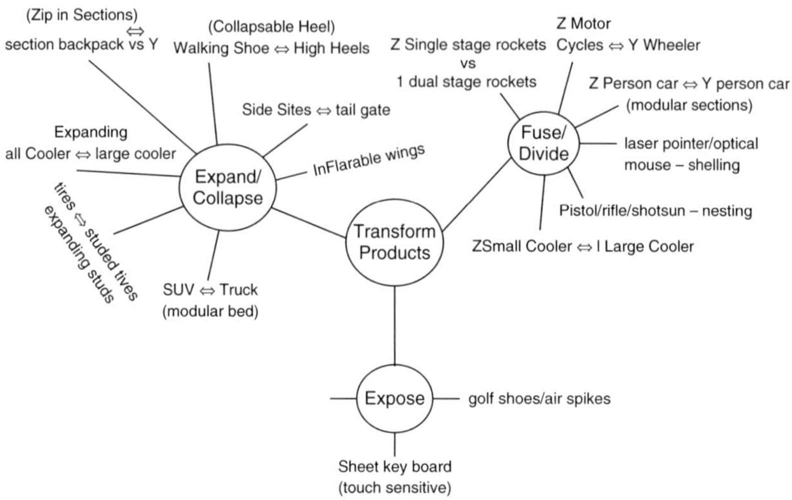

Figure 9–3 Example mind map from pilot study with graduate student participants.

While the principles and facilitators provide foundations for a transformational design theory, a question integral to successful design of transforming systems yet lingers—When do we need a transforming system? Transforming systems are time-dependent, in that one state and its function cannot be used simultaneously with the other state(s) and function(s). In other words, transformation should not be pursued if both configurations are needed at the same time. Through the inductive approach of studying existing transforming systems and by hypothesizing results of that induction, the following indicators hold true in transformational systems. Systems that appear to be "ripe" for implementation as transformers are:

- Systems needing packaging for portability and deployment,
- Multiple systems allowing consolidation into one system for convenience and the efficient use of resources, and
- Multiple systems having dissimilar configurations sharing common material and/or energy flow

These indicators give a first glimpse of when transformation may be beneficial. Current research is being conducted to study the correlations between the transformation indicators and the functionality of a device. The main point of these indicators is that, while the savings in volume, weight, and portability may be the most obvious advantage of transformers, there exist usage situations where a functional metamorphosis provides a greatly improved candidate solution to a design problem. The indicators are a first step to analytically determining these situations. Although the principles, facilitators, and indicators provide new understanding of the development of transforming systems, additional design guidance is needed in pursuing transformation solutions to novel or common design problems. This additional guidance, as a first incarnation, is provided in our Transformational Methodology.

Transformation Methodology

Within the context of the principles for transformation given above, the desire to incorporate a methodological approach emerges. This section describes such an approach.

Hierarchical (or Categorical) Approach to Design

The conventional approach in defining a problem and gathering needs and requirements has been reconditioned. The hierarchical approach (Fig. 9–4) explained in this section takes the current problem or need and creates an abstract problem scenario, or a *Generalized Scenario.* From this scenario, predicted or anticipated uses of the system, *Objectives,* are extracted. *Customer Needs* are then gathered from each objective to create a comprehensive list of needs across the *Generalized Scenario.* From this set of needs, high-process-order solutions are created. These are termed *Capabilities* and give a first-level insight into effective solutions to the needs relating to *Objectives* and to the *Generalized Scenario.* This type of approach not only helps capture various possible, present, or future needs during the design of a system, but can also help designers at a managerial level decide the outcome of their design by scrutinizing a bigger picture of the problem. The following sections explain this hierarchical, or categorical, approach.

Understanding a Generalized Scenario

Generalized Scenario—An abstract statement describing the overarching extent of the problem. For example, "a system for surveying and defending a large open area" may be used as a generalized scenario. The idea of creating

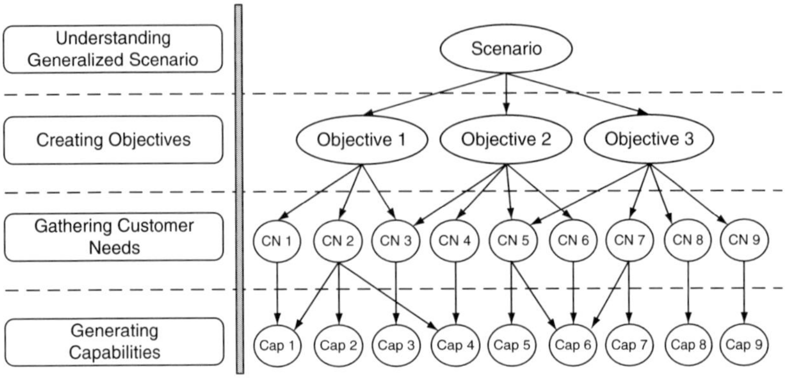

Figure 9–4 Hierarchical approach flowchart.

a generalized scenario is to take a step back, analyze a problem and enter the space of possible uses (current or future) of the system being designed. Stating a general scenario in the context of a specific usage of the system not only captures that usage but helps in anticipating and predicting other existing or not-yet-existent uses of the system. This activity not only helps gather *Objectives* for the system but encourages a designer to anticipate and think about non-obvious needs and future needs.

Creating Objectives

Objective—An anticipated event or sequence of events projecting the planned or possible uses of a system in the context of the generalized scenario. For example, "Survey pipeline in the desert," "Interrogate prison inmates in specified perimeter," and "Defend designated area of thick foliage" could be objectives for the previously stated generalized scenario. *Objectives* are more specific descriptions of what the system must do, but are not a fully refined list of *Customer Needs*; they broadly define what the system must do in the context of the *Generalized Scenario.*

Gathering Customer Needs

Customer Need—Requirement of the system stated in the context of an objective. There are general categories into which customer needs can be grouped to understand their differences. For example a need could be to "Survey area stealthily" or "Travel through different weather conditions." By gathering needs for each objective individually, a more comprehensive set of needs is created that now captures the goals of the system expressed in the objectives and generalized scenario. The next step is to generate solutions to these needs that are not form-specific, maintaining abstraction. These abstract solutions (identified as "capabilities" below) aid in developing a broad design space of form-specific solutions for the next stages of the design process.

Generating Capabilities

Capability—A high-order process-oriented task enabling a customer need or set of customer needs. This task is not form- or technique-specific. For example, given the customer need above, "survey area stealthily," a possible capability may be to "hover." We can then embody this capability in the system by allowing for the system to hover using gases, rotors, jet engines, magnetic levitation, etc. A single capability may or may not relate to more than one customer need. In this case "hover" does relate to the needs of "survey area stealthily" and "travel through different weather conditions." However, a capability like "perch" relates to "survey area stealthily" and not to "travel through different weather conditions," as "perch" isn't catering to the need of traveling.

State Extraction

The purpose of the product hierarchy is to equip the design engineer with a plethora of information pertaining to the essentials of the design (expressed in the objectives and customer needs) along with the general means to satisfy these needs (expressed in the capabilities). Not only does this process force the designer to contemplate the nature of the design problem, this process causes engineers to state their thoughts, insights, and creative avenues in tangible form. The cohesion of this information represents usage knowledge—one of the greatest tools a design engineer can possess.

As with any design problem, the final goal is to provide an innovative, quality product that satisfies the comprehensive list of needs expressed by the customer. The first step in materializing a product from the process outlined thus far is state extraction. The development of states directly corresponds to the previously generated set of capabilities. Inasmuch as states are spawned from capabilities, a state can also be considered a specific physical embodiment of a capability. As an example, the capability to *fly* has several states, including propeller-driven airplane, jet airplane, helicopter, rocket, ornithopter, and flying saucer.

Rehashing the design process to this point, the designer starts with a general scenario, from which objectives are created. For each objective, customer needs are gathered, representing the requirements for successful execution of the objective. Next, capabilities stem from the needs as high-level process solutions to carry out the respective needs. And from the capabilities, we generate state solutions, or more specific and physical forms of a capability. The next step in the transformational design method is to provide methods, building upon the transformation principles and facilitators, to generate these solutions.

Transformation Route of Design

The steps outlined thus far simply provide a systematic mode of obtaining as much knowledge as possible regarding the design problem. With this in mind, the designer must take a step back and reexamine the big picture.

In an ideal situation, a product satisfies the entirety of objectives and customer needs. However, products rarely accomplish this lofty goal. It is common, for example, to have conflicting customer needs. In the context of automobiles, for example, customers want high performance ratings yet also high fuel efficiency. While the need for high performance does not necessarily oppose high fuel efficiency, it is actually at the state level where the conflict resides. Elaborating on the automotive example, a high performance automobile may have a rather large engine supported by a large frame. On the other hand, a fuel-efficient car normally has a smaller engine and a lighter frame. The results of such conflicts often result in a compromise where neither of the needs is satisfied holistically but each need is satisfied with some compromise. The goal of any design engineer, however, is not just to satisfy the needs at a sufficient level but to completely and absolutely satisfy the customer. This goal is where innovation applies.

The ability to solve the totality of customer needs, even the conflicting or contradicting needs, is a paradigm shift from more conventional design theory. Transformers may provide new insight and solutions here. The purpose of a transformational product is to be able to execute an objective requiring or dependent upon a certain state and then transform to a different state in order to fulfill a different objective. Referring the hierarchical breakdown of product usage in Figure 9–4, the designer should explore transformation when encountering the situation where different objectives requiring independent states are necessary to carry out the general scenario. This heuristic for transformation is general. Its implementation must be supported by ideation techniques that assist the designer with categories and mental cues for retrieving or searching for analogical solutions.

Currently, there is no complete systematic tool to quantitatively show the relationships between the steps in the transformational design process

(Generalized Scenario → Objectives → Customer Needs → Capabilities → States). In previous publications, we reported progress in using a "design matrix" approach to describe the relationships between these steps (Skiles, 2007). We are continuing this research and plan to report the results in future publications.

Ideation for Transformation: Tools for Innovation

With the development of a scenario, objectives, and customer needs of the design problem must be related to capabilities and state extraction, concept generation techniques discussed in this section help in further embodying transformation. To aid in the generation of transforming concepts, we employ transformation design principles and facilitators in an extended mind-mapping technique and in the form of innovation cards. While conventional concept-generation techniques can assist in the development of a transforming product, the transformation principles and facilitators act as a directed tool to enable efficient contrivance of transformers. Existing ideation methods such as mind-mapping and brain-writing can be used in conjunction with these principles to generate ideas for transformation. These techniques are discussed below.

Extended Mind-mapping

The traditional mind-mapping approach is to write the problem to be solved in the center of a black sheet with a box around it. Ideas are generated to solve the central problem and are recorded in branches from the problem statement. As ideas are refined or spawn other ideas, these are connected to the parent idea on the map through category descriptors. These categories are more abstract and higher-level solutions that provide mental cues for specific ideas (Otto and Wood, 2001).

This technique is adapted to aid in the generation of transformers. The basic process is the same, with the transformational design problem in the center of the map. The problem is stated in the form of the two (or more) objectives of the transformer, for example *Store / Fly* in Figure 9–5. The designer then chooses design principles and facilitators that may be of use in the development of a transition between the states and places these as branches around the problem statement. Ideas are then generated that are specific to each principle and connected as branches. As with a traditional mind map, each new idea can grow new branches of its own. Special attention should be paid to interactions between the ideas attached to different principles, since transformers frequently arise from a combination of different principles and facilitators.

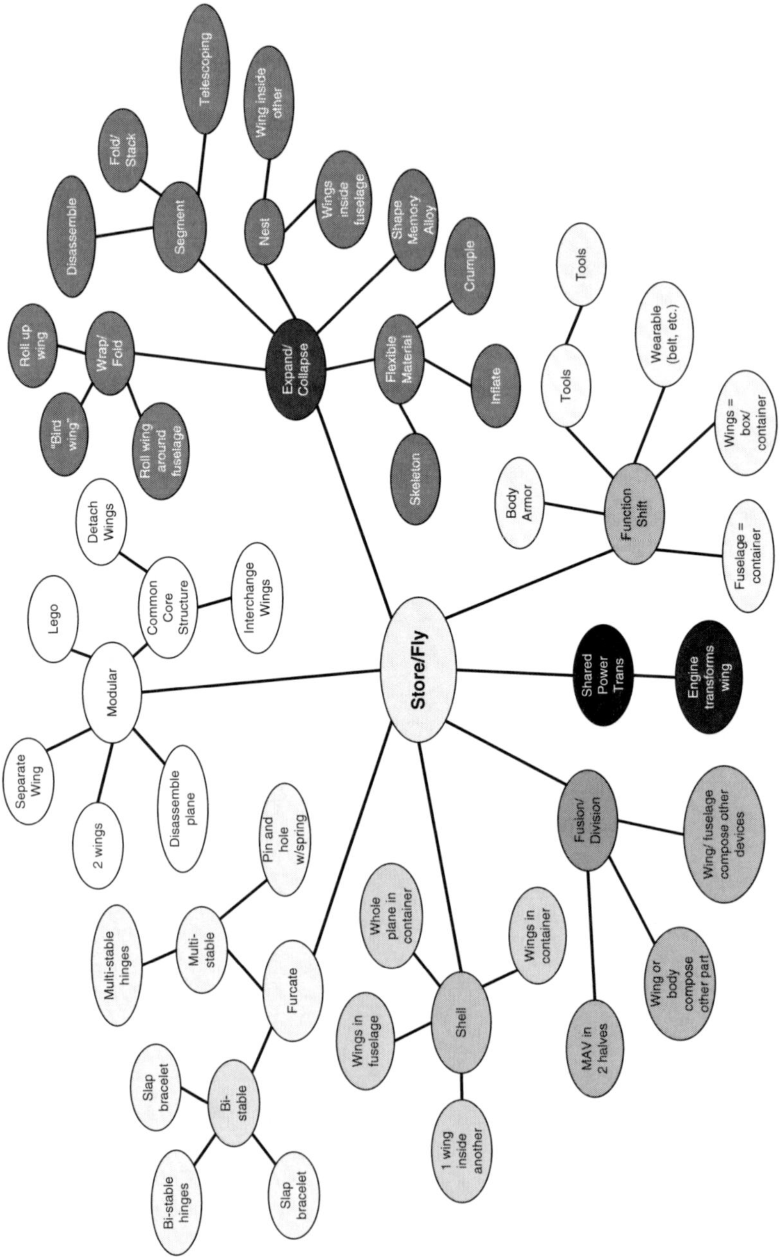

Figure 9–5 Extended mind map of the states "Store / Fly".

Mind maps can be used in specific design problems. For example, consider the problem of providing a screwdriver with different head configurations. The mind-map would contain transformer principles and facilitators that may be combined to direct designers to innovative solutions, such as the folding screwdriver, the fuse and divide screwdriver, the expand and collapse screwdriver, or the multi-principle screwdriver.

Transformation Cards: T-Cards

As an alternative and complementary method, we have created a set of "T-cards" to be used in the concept-generation process. Each card shows one of the transformation principles or facilitators along with examples (general analogies) of how the principle/facilitator is embodied. This deployment brings the design principles and facilitators to the designer in a simple yet creative environment. The transformation cards are 4" x 6" and coded with color and geometric shapes. The color and geometric codes relate a principle to its facilitators. These relationships exist because certain facilitators have been found to aid a certain type of transformation captured by a transformation principle. For example, a facilitator such as *Shared Power Transmission* does not facilitate the principle *Expand/Collapse* in its embodiment (based in the inductive and deductive research), but it will facilitate the principle *Fuse/Divide*. The alignment of color pathways between the T-cards provides these relationships.

The transformation cards are used in two primary ways for concept generation. First, the cards can be used sequentially. For example, one transformation principle is selected, and different combinations of transformation facilitators under that principle are considered to apply a form of transformation. This approach can also be used in reverse, starting with a facilitator. Using the geometric and color codes, a facilitator is linked to other facilitators, and ultimately linked to a principle to generate a transformation solution. For example, consider the problem of storing a Micro Unmanned Aerial Vehicle (MAV). Historically, MAVs have been stored in either a disassembled state that required a series of assembly operations or in an assembled state that required a relatively large space. The wing usually requires the greatest storage volume per unit mass of the plane and therefore presents the greatest challenge when attempting to reduce its stored volume. The Air Force Research Labs have developed the Tactical MAV (TACMAV), which addresses this problem by building the wings from flexible carbon fiber so that the wings can be rolled into a container for ease of storage and portability. The 50 cm–long TACMAV (53 cm wingspan) uses *flexible* (*Material Flexibility facilitator)* wings that can be *rolled* (*Roll/Wrap/Coil facilitator*) around its fuselage, allowing it to *collapse* (*Expand/Collapse principle*) and be stored in a 13 cm–diameter tube carried in a soldier's backpack. When pulled out of the tube, the folded wings automatically snap into place (*Furcation facilitator*). The cards are able to capture this embodiment. Indeed, use of these cards led to a solution that reduced the storage volume by 40% over previous designs.

Direct Design by Analogy

Another way to use the cards for concept generation is to facilitate design by analogy through current or historical devices. *Analogy* is defined as "a similarity between like features of two things, on which a comparison may be based." Concept generation often involves use of analogy in an implicit fashion. Research shows that a more explicit version of design-by-analogy will dramatically increase the number and novelty of solutions generated (Linsey, Murphy, Markman, et al., 2006). The chapters in this book by Christensen and Schunn (Chapter 3), and by Markman, Wood, Linsey, Murphy, and Laux (Chapter 5), provide additional insights into the use and promotion of analogical reasoning for innovation.

Active research is studying specifically how analogies should be incorporated into the concept generation process in order to maximize production of innovative solutions. The design principles governing transformation help generate a form-specific solution to the design problem requiring transformation at the systems level. The use of T-cards is one way of explicitly using analogies through pictures of existing and historical devices. All the cards are laid out in front of the designer(s), which sparks new ideas by creating an atmosphere of analogies that the designer can pick from at random and extract analogous solutions. The designer can randomly select card(s) and then apply a hierarchical approach to create more transformation embodiments. Similar to how the transformation principles and facilitators are created (inductive process), analogies can be found in biology: micro- and cellular level, zoology, plant biology, human anatomy and associated mechanics; in physics: state changes, quantum mechanics, relativity, classical mechanics; in chemistry; and in current systems: patent searches, consumer products, manufacturing systems and techniques, etc.

Application of the Transformational Design Method

In order to demonstrate the effectiveness of the proposed methodology, designers applied the technique to develop a cycling accessory. First, the designers decided on a general scenario: Take a day-long trip or commute on a bicycle. Then they developed five objectives based upon the general activity of riding a bike: Secure the bike, perform maintenance, transport cargo, ride in different environments, and carry personal items. The objectives do not directly pertain to the act of riding a bike but are important supporting processes that are common occurrences corresponding to the overall cycling activity.

The second step of the design involved the designers' generating a comprehensive list of customer needs for each objective. In order to successfully execute the objective to *secure the bike*, the designers determined that the device had to exhibit the following qualities: *be tamper-resistant, weather-resistant, and have quick and easy locking and unlocking procedures*, etc. To successfully complete the objective to *perform maintenance*, the device needed the following traits: *Store tools on bike, know the tire pressure, and exert minimal human effort*, etc. An abundance of needs for each objective were developed but only a portion is provided here. Refer to Figure 9–6 for a condensed version of the methodology results.

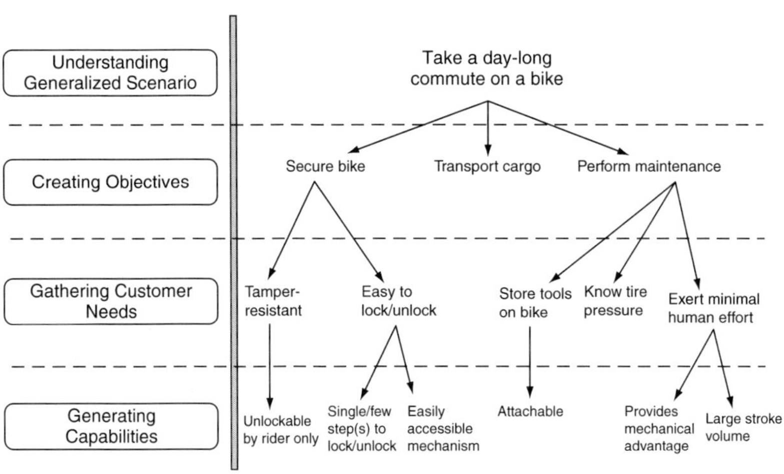

Figure 9–6 Hierarchical approach applied to a scenario.

Subsequent to constructing a list of customer needs, the designers produced capabilities for each need. To effectively generate capabilities, the designers contemplated ways to accommodate or solve each need. Another way to describe a capability is the manner in which one facilitates a particular need. For example, the designers listed the capabilities of *single/few step(s) to unlock/lock* and *easily accessible locking mechanism* (and others not mentioned here) to solve or accommodate the customer need of *locking the bike*. The next step involves analyzing the lists of objectives, customer needs, and capabilities to unveil insights regarding the design.

For this application, the designers considered all the information gained from the list of objectives, needs, and capabilities, then subjectively chose the most relevant needs and capabilities to determine the objectives most likely to facilitate a transformational design solution. These objectives were *secure bike* and *perform maintenance.*

By considering these two objectives along with the key customer needs and capabilities, they derived two states (a single state per objective). The first state, a U-shaped bike lock, relates to the objective to *secure the bike.* The customer needs of *easy to lock/unlock, tamper-resistant, weather-resistant, stored on bike easily,* and others further directed the concept generation. Lastly, the capabilities of *waterproof, unlockable by rider only, attachable, single/few step(s) to lock/unlock,* and others further guided the students in state visualization. The second state, a hand-actuated air pump, relates to the objective to *perform maintenance.* The customer needs of *store tools on bike, know tire pressure, exert minimal human effort, nozzle should fit valve easily, stores on bike easily,* and others assisted concept generation. Furthermore, the capabilities of *indicate exact pressure, large stroke volume, provides mechanical advantage, flexible nozzle,* and others further directed the designers in the extraction of this state.

With the two separate states known, the next thought process was to determine whether transformation should be pursued. The two objectives require separate states that are not used simultaneously, hinting that transformation is a promising avenue. Furthermore, this application fits the convenience transformation directive in that the two systems having individual configurations allow consolidation into one system for convenience. After reaching this key milestone in the transformation design process, the designers began the detailed concept generation process.

Transformation cards developed for this phase of design were used in two ways to generate concepts. In one ideation activity, direct design by analogy was used, where the principle and facilitator cards were spread out to inspire analogous transformation solutions.

A second concept-generation activity was carried out where the embedded relationship between principles and facilitators, as dictated by the codes in the T-cards, was used. This activity produced concepts already captured in the design by analogy exercise but also yielded some unique and non-obvious concepts not seen from the previous technique. Concept 8, which was generated from the T-Card activity, embodies a simple and novel transformation solution. As T-Cards capture a sequential approach in using the transformation principles, where certain facilitators aid different principles to create transformation, the activity provided more insights by generating various combinations of transformation embodiments.

All three principles, *expose/cover, fuse/divide* and *expand/collapse,* proved to be extremely helpful in generating this concept. A solid model of the transforming bike lock and tire pump involves the principle of *expand/ collapse*, where the facilitators *wrap/fold* and *material flexibility* lead to the use of a flexible hose that could be wrapped for storage. Furthermore, it was thought that the wrapped hose could be stored in the device by incorporating the *shelling* facilitator. This facilitator also falls under the *expose/cover* principle. The *expose/cover* principle suggested exposing an unused space in which the hose can be stored in the lock configuration. The principle of *fuse/ divide,* with relation to the *segmentation* and *function sharing* facilitators, aided the designers in converting the U-section of the lock into pumping handles and as a means for guiding the reciprocating motion of the pump. An analogy to the way the device is used in its pump configurations is a pair of pruning shears. The location of the pumping mechanism is housed inside the end component, which is integrated with the locking mechanism. To develop this idea, the designers simultaneously applied the *shelling* and *function sharing* facilitators.

This device will successfully accomplish the objectives to *secure the bike from theft* and *perform maintenance* by embodying such a design. For example, to accommodate the capabilities of *large stroke volume* and *provide mechanical advantage*, the piston-cylinder assembly is adjustable along the length of the U-section. This allows for variation in stroke volume and also in moment-arm length.

The design was pursued through to the prototyping phase. The first fully functional prototype demonstrates the feasibility and manufacturability of the design. For ease of manufacturing, most components were specified to correlate to available common stock sizes. The cylinder was constructed of aluminum tubing. The U-section was constructed of stainless steel rod. The remaining components were constructed of aluminum and steel, except the piston and top cap, which were created with high-density polyethylene and

Teflon, respectively. For this prototype, commercial detents secured the chamber assembly in each configuration. The next iteration would involve optimizing the chamber dimensions to fine-tune the pressure and volume capabilities, along with further developing the locking mechanism and jointed section for maximum security.

Conclusion

Transforming products have tremendous potential benefits in a wide array of applications. The benefit comes from their ability to change state and facilitate new functionality; all within a single system. This research leads to a theory of transformation encapsulated in a set of transformation principles and facilitators. These form a basis from which a transformational design methodology is developed. The ultimate goal is to have a repeatable method, not only to reveal the opportunity for transformational devices, but also to deploy the theory and physically embody transforming products that have abilities unparalleled by any other product architecture. This chapter presents the developments in the current progression of the transformation design theory by describing a renewed approach for generating and analyzing system usage scenarios, objectives, customer needs, and capabilities. The method provides an avenue for developing transforming systems. The initial stages of the method are followed by concept generation techniques that use the transformation design principles and facilitators.

For such a methodology to be widely accepted and repeatable, the framework of the methodology may be refined for consistency, simplicity, and accuracy when applied to a wide variety of design problems. The next major area of emphasis lies in device functionality. Further, mathematical tools for transformational analysis and state extraction, and more concept generation tools and techniques, are currently being explored. These future improvements will help us consistently design and embody new, innovative products using transformational solutions.

The design principles highlighted in this chapter are a means of innovation. They are, in essence, meta-analogies that provide mental cues, in concert with ideation innovation tools, from which a wide array of analogical solutions is possible. The aim of these principles, and, more generally our research model, is to develop innovative solutions to difficult problems. Our world is a place of constant change. To compensate for or accommodate this change, designers must continually seek innovation.

Acknowledgments

The authors would like to acknowledge the support provided from the Cullen Endowed Professorship in Engineering, the University of Texas at Austin, and grants from the Air Force Research Laboratory Munitions Directorate (AFRL/MN) at Eglin, Florida, the Air Force Office of Scientific Research (AFOSR), and the National Science Foundation under Grant No. CMMI-0555851. The authors would also like to thank the Department of Engineering Mechanics at the U.S. Air Force Academy for their support and guidance. Any opinions, findings, or conclusions found in this chapter are those of the authors and do not necessarily reflect the views of the sponsors.

References

Abdulrahim, M., Garcia, H., and Lind, R. (2005). Flight characteristics of shaping the membrane wing of a micro air vehicle. In *Journal of Aircraft*, Vol. 41, No.1, January–February, pp. 131–137.

Altshuller, G. S. (1984). *Creativity as an exact science.* Luxembourg: Gordon and Breach.

Baldwin, T., Rodarte, L. E. E., Karidis, J. P., and Moffatt, S. S. (2004). Folding keyboard. U.S. Patent Number 6793421.

Boothroyd, G., Dewhurst, P., and Knight, W. (2002). *Product design for manufacture and assembly*, 2nd ed. New York: Marcel Dekker.

Colledge, A. L., and Johnson, H. I. (1989). Portable multi-purpose exercise device. U.S. Patent Number 4,856,775.

Ericsson, A., and Erixon, G. (1999). *Controlling design variants: Modular product platforms* (p. 145). New York: ASME Press.

Geary, J. A. (2000). Geary convertible crutch system. U.S. Patent Number 6,085,766.

Green, M. G. (2005). Enabling design in frontier contexts: A contextual needs assessment method with humanitarian applications. Ph.D. dissertation, University of Texas, Austin.

Linsey, J. S., Murphy, J. T., Markman, A. B., Wood, K. L., and Kurtoglu, T. (2006). Representing analogies: Increasing the probability of innovation. In *International Design Engineering Technical Conferences*, Philadelphia, Pa., 10–13 Sept., DETC–2006–99383.

Mollerup, P. (2001). *Collapsible: The genius of space saving.* San Francisco: Chronicle Books.

Otto, K., and Wood, K. (2001). *Product design: Techniques in reverse engineering and new product development.* Upper Saddle River, N.J.: Prentice Hall.

Pahl, G., and Beitz, W. (1999). *Engineering design*, 2nd ed. London: Springer-Verlag, Ltd.

Pugh, S. (1996). *Total design: Integrated methods for successful product engineering.* T.J. Press, Cornwall, Great Britain.

Singh, V., Skiles, S. M., Krager, J. E., Wood, K. L., Jensen, D., and Szmerekovsky, A. (2006). Innovations in design through transformation: A fundamental study of transformation principles. *International Design Engineering Technical Conferences*, Philadelphia, Pa., 10–13 Sept., DETC–2006–99575.

Singh, V., Warren, L., Putnam, N., Walther, B., Becker, P., Danielson, A., Koraishy, B., Wood, W., Jensen, D., and Szmerekovsky, A. (2006). A novel exploration into gust-resistant operation of MAVs / UAVs through transformation. 2nd US-Euro MAV Conference, Destin, Fla.

Skiles, S. M. (2006). Development of principles and facilitators for transformational product design. Master's thesis, University of Texas, Austin.

Skiles, S. M., Singh, V., Krager, J. E., Seepersad, C. C., Wood, K. L., Jensen, D., (2006). Adapted concept generation and computational techniques for the application of a transformer design theory. *International Design Engineering Technical Conferences,* Philadelphia, Pa., 10–13 Sept., DETC–2006–99584.

Skiles, Stewart (2007). Development of principles and facilitators for transformational product design. Master's thesis, University of Texas, Department of Mechanical Engineering, Spring.

Sushkov, V., Mars, N., and Wognum, P. (1995). Introduction to TIPS: Theory for creative design. In *Journal of AI Engineering*, Vol. 9, No. 3, pp. 177–189.

Tennant, L. H., Herrin, A. A., Simmons, G. L. (1988). Towel that converts into a bag, U.S. Patent Number 4,794,029.

The Unmanned Switchblade. Retrieved on 4/27/2007 from http://www.popsci. com/popsci/aviationspace/0f2505a52aceb010vgnvcm1000004eecbccdrcrd.html.

CHAPTER 10

INTRODUCTION OF DESIGN ENABLING TOOLS

DEVELOPMENT, VALIDATION, AND LESSONS LEARNED

JOSHUA D. SUMMERS

SRINIVASAN ANANDAN

SUDHAKAR TEEGAVARAPU

THE future of product realization will be driven by increasing complexity and globalization. Designers are challenging the limits of design to create innovative products that are distinctive and well received in the marketplace (Schunn et al., 2006). We believe that design and innovation are not two separate activities. While all designs may not be innovative, all innovations are certainly designs of some kind. Furthermore, there is an element of uncertainty in terms of repeatability of a design process and associated methods; this uncertainty is more widespread if the aim is to achieve innovative designs. As the argument over definitions and distinctions between "design" and "innovation" continues in the community, it is wise to focus our efforts on developing robust processes, methods, and tools for design. These processes can be extended for "innovation," since it is believed that innovation is a byproduct, albeit desirable, of design.

To provide agility to engineers in this challenging environment, design must move beyond using computers only for engineering analysis and document archival. Imagine a future where, to overcome these new challenges, a designer collaborates on a project with other engineers across the world and with computer design agents that activate themselves as needed. Further imagine that this designer has the ability to create new design-enabling computer agents without writing computer software, while being able to *predict how introducing these design enablers will modify the process*. This is the future of engineering design envisioned by the researchers in the Automation in Design (AID) Group at Clemson University: computers and humans truly collaborating, where engineers can build their own design enablers in a natural manner and informed by how design enablers will affect design processes. To achieve this vision, two research paths underlie the long-term research plan of AID: (1) a path developing a theory of collaborative design, and (2) a path researching and modeling the impact of design enablers in design. These paths agree with the strategic planning report from the 2004 NSF Engineering Design workshop, where two of the three specific thrusts identified were (1) social-technical aspects and (2) design informatics (Shah et al., 2004).

While both paths are followed, the latter path, design enabler research, is the focus of this paper. In this paper, we will define *design enablers*, giving examples of these as developed in the AID Group, and discuss lessons learned in the development and validation of these tools. The aim of this paper is to provide a foundation for other researchers to begin to work together in a more structured and systematic manner in researching design engineering while meeting the demands and needs of our industry constituents.

We loosely define *design enablers* as tools that are found in the design process, both computational and non-automated, that enable design engineers in the product realization process. These tools can range from problem-definition tools such as Quality Functional Deployment (QFD) (Cohen, 1995), Problem Definition and Specification (PDS) (Ulrich and Eppinger, 2004), to idea generation tools such as collaborative sketching (C-Sketch) (Shah et al., 2001), morphological matrix (Zwicky, 1969), Theory of Inventive Problem Solving (TRIZ) (Altshuller, 1999), and from reverse engineering such as Subtract and Operate (SOP) (Otto and Wood, 2001), to optimization such as Analytic Target Cascading (ATC) (Kim et al., 2002). With this admittedly wide-ranging definition, we cast a large net intentionally.

Each of these tools has different origins; they were developed with different motivations and grounded in different levels of formalism. For example, QFD was developed through years of observation and refinement in industry where the customer's definition of quality must be translated into engineering targets and metrics (Cohen, 1995). In contrast, C-Sketch was developed at the Design Automation Lab at Arizona State University in an attempt to refine existing idea-generation tools such as Gallery Method and Method 6-3-5 to bring together their respective strengths (visual representation and provocative stimuli through misinterpretation) in a single tool. Because each tool was developed with different objectives, their respective evaluations and validation are also distinct.

We argue that there are different scenarios leading to the development of design enablers; two major categories of which are (1) demand driven and (2) internally derived.

1. Demand driven scenarios are those in which the goal of the research is to develop new tools for specific challenges.
    a. One example of "push" from academia to industry is where tools that are developed in academic research, such as the design exemplar toolset, are customized for specific industrial applications, such as a Michelin lamelle search and retrieval tool (Summers, Divekar, and Anandan, 2006).
    b. Likewise, industry may "pull" these tools from academia, requesting the development of new design enablers for specific uses, such as the development of a requirements modeling tool for design trade-off scenarios developed at AID for BMW (Mocko et al., 2007), (Maier et al., 2007).
    c. A third class of demand-driven design enabler tool development may be anticipatory development of tools where industry has not yet been directly involved in the design enabler tool development. An example of this might be the original development of the design exemplar (Summers, Shah, and Bettig, 2004). The distinguishing characteristic of this category is that the tools are the focus of the research.
2. The second category of design enabler tool development includes internally derived tools: tools that result from the experiences of the designer as an attempt to improve the design process. Specifically, these tools, such as the connectivity graphs and reverse failure modes effects analysis (RFMEA), are not the primary objectives of the industry sponsored research, but are by-products of the design work (Snider and Summers, 2006), (Snider et al., 2006).

This paper will look at three specific cases for design enablers that have been developed in the AID group: the design exemplar-enabled lamelle retrieval system, the requirements modeling concept-exploration tool, and the connectivity graphs. The objective is to highlight the differences between the evolution of these tools and how this impacts their validation, implementation, and assimilation. The ultimate goal is to be able to determine the impacts that these tools have on the design process, while the first step and the motivation of this paper is to more fully understand the development of these tools.

Design Enablers at AID

The Automation in Design Group has developed several new design enabling tools over the past several years (see Table 10–1). Three of these tools are discussed in this chapter to illustrate how each of these tools has evolved and how each has been validated. The approaches to validation have included case studies (both industry-focused and problem-focused), pilot user studies, and intrinsic validation. *Intrinsic validation* is where the individual components and characteristics of a design enabler are examined based on previous known performance as reported in the literature.

Design Exemplar-Enabled Lamelle Retrieval System

The design exemplar was originally developed to model geometric and parametric design problems. In standard mechanical component design, such as gear trains, belt drives, transmissions, or beam structures, parameters can drive form design even as the shape simultaneously influences different design parameters. Hence the design exemplar was introduced as an integrated approach to representing design problems. A set of entities and constraints has been offered for representing semantic, topological, geometric, and algebraic relationships found in mechanical engineering design. These entities and constraints are incorporated into bi-partite graphs for representing the design models, thus integrating both parametric design and geometric design. The bi-partite graph is partitioned in two ways: match/extract partition (used for validation) and alpha/alpha_beta/beta partition (used for transformation). Design exemplars have been used

TABLE 10–1 Design Enablers at AID.

Name of Tool	Type of Development	Type of Validation	Publication
Agent Negotiated Fixture Design via Common Representation	1.c	Intrinsic, Demonstration	[16–19]
CAD Query Language	1.c	Intrinsic, Demonstration	[10, 20, 21]
Resistance Based Collaborative Design Model	1.c	Intrinsic, Demonstration	[22, 23]
PRSM: Proper Review Selection Matrix	2	Intrinsic, Case Study (Industry)	[24, 25]
Systematic Pruning of Morph Charts	1.c	User Study	[26]
Reverse Failure Modes Effects Analysis	2	Case Study (Industry)	[15, 27]
Function-Component Structure and Algorithms	2	Case Study (Industry)	[15, 27]
Interrelation Matrix	2	Case Study (Industry)	[15, 27]
Connectivity Graph	2	Case Study (Industry)	[15, 27]
Reverse Engineering Database	2	Case Study (Industry), Case Study (Academic)	[14, 27]
Lamelle Retrieval System	1.a	Case Study (Industry), Case Study (Academic), User Study (Customer)	[10, 28]
Exemplar Production Systems	1.c	Intrinsic, Demonstration	[29, 30]
Concept Selection for Varying Levels of Abstraction	2	Case Study (Industry), Case Study (Academic), Intrinsic	[31]
Requirements Modeling Matrices	1.b	Case Study (Industry), Case Study (Academic), User Study	[11, 12, 32]
Frame Design and Analysis Tool	1.b	Case Study (Industry), Intrinsic	[33]
Rule-Based Design Guidelines and Protocols	1.b	Case Study (Industry)	[34]

in feature-based design and feature recognition systems, used to model standard design procedures, used for rule validation and querying. The design exemplar has been investigated as a CAD query language by comparing the components of the de-facto query language, SQL, with those of the exemplar. For a detailed explanation of the design exemplar and its use as a CAD query language, the reader is referred to Summers, et al. (2006) (Summers, Divekar, and Anandan, 2006).

The exemplar technology was presented to a large tire manufacturing firm in North America. On explaining the concepts and uses of the design exemplar, it was discovered that this technology could be used to solve one of the firm's design problems. The company designs mold inserts for its tire treads. The tooling cost for each mold insert is approximately $2,000. Hence, in order to reduce the manufacturing cost, the designers would like to retrieve inserts that are similar to the desired shape geometry, so that the same tooling can be used. An example of such a target mold insert is shown in Figure 10–1.

The designers wanted to find all such inserts that fall within a tolerance envelope. For example, the lamelle in Figure 10–2(a) can be considered to be a target model. The tolerance envelope is indicated by the region in red. The lamelle in Figure 10–2(b) is an example of a mold insert that fits inside the tolerance envelope.

The industry's ideal would be for the designers to manually search through the whole database of lamelle models to find similar sets. As this exercise is tedious and not practiced, it formed the motivation for the introduction of design exemplars as a CAD query language to facilitate automatic lamelle retrieval. The usefulness of the exemplar was validated through case studies in the form of testing against a sample database of

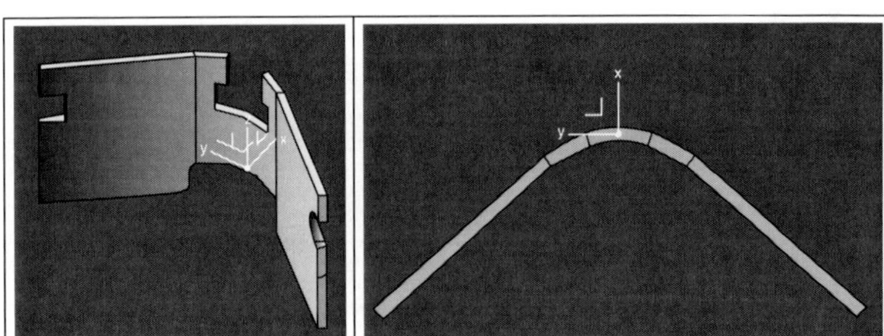

Figure 10–1 Example of lamelles found in a database for retrieval.

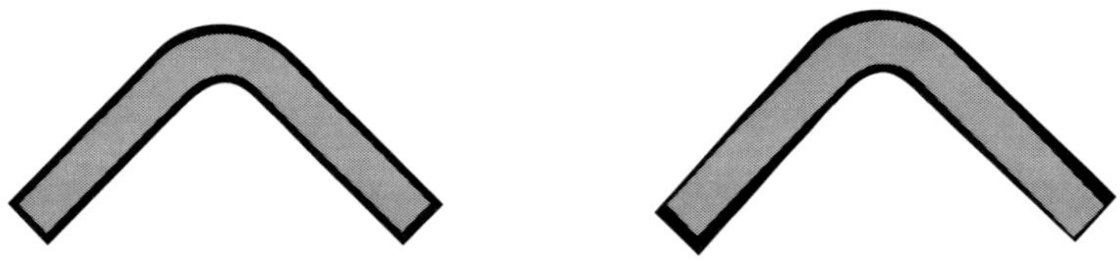

Figure 10–2 Envelopes for lamelle similarity retrieval.

lamelles. The results obtained from actually running the query against the lamelles in the database were compared to the theoretically expected results.

There were certain limitations faced while implementing the proposed solution. The major hurdle was understanding the problem as presented by the customer. The complexity of the problem was not well understood by either the industry sponsor or the research team before the solution development started. Specifically, the needs that were originally identified were not the actual needs; the customer did not need the full functionality of defining geometric queries that is supported with the design exemplar system. Rather, the customer needed a way to query the same geometry repeatedly with slight variations in the parametric values. The vocabulary of the query set was a much smaller subset of the query language vocabulary. Through testing, iteration, and prototyping of the envisioned system, a better understanding of how the customer would use the design tool led to the eventual conclusion that the exemplar technology was much more advanced and robust than was actually needed. There were other limitations, such as licensing for the commercial software that the company was using. Also, the exemplar technology was implemented in an academic environment where transferring the technology to a commercial environment was challenging, given the export-control issues.

This tool has been, and is being, validated through customer use, where customer acceptance is the measure of its worth. Detailed experimentation to determine tool effectiveness in performing the desired tasks is not required by this sponsoring organization, nor is a comprehensive "before and after" case study. However, researchers in the AID Group have conducted a localized case study to model the design process that includes lamelle design in the general tire tread design process. This case study will be supplemented with a second one studying the design activities after the introduction of the design enabler. It is important to take two process "snapshots" before and after the introduction of the design enabling tool to determine whether the tool actually changes the activities or whether it just improves their efficiencies. These two case studies were included as a part of the research agreement with the sponsor so that AID investigators can have at least limited access to the designers after tool implementation and acceptance. This agreement is critical to the success of design enabler development research programs as joint ventures between industry and academia.

This approach of pushing technology into industry in the development of design tools, or design enablers, has both advantages and disadvantages. First, a major advantage is a real-world assessment of the academic technologies that have been developed with the goal of aiding the design process, specifically the design exemplar. Second, in this interaction with industry designers, researchers have access to "expert" designers who are not readily available in typical academic settings. This creates opportunities to study actual design processes in industry to explore how new tool introduction can impact the process.

Requirements Modeling for Concept Trade-off Exploration

A major automotive OEM (original equipment manufacturer) approached the AID group with a request to develop a method or a design enabler to support the evaluation of early-stage vehicle concepts with respect to lightweight engineering. A secondary objective of this tool was to allow designers to trace requirements through functions and components to the specific tests that are performed, thus extracting which tests are worst-case scenarios and which tests are redundant in validating the requirements. In this case, the design enabler was not based on any existing technology, nor was it clear at the beginning of the project what form the tool would take when it was finished.

Specifically, this tool is a matrix-based mapping model between seven different design domains: requirements, functions, working principles, components, component parameters, test parameters, and tests. The tool provides a systematic set of algorithms to populate the matrices based on existing products, such as automotive cooling systems, accelerator pedals, and car seats. Additionally, algorithms are derived to manipulate the matrices to explore design trade-offs for different configurations, such as replacing, modifying, deleting, adding, and combining components, working principles, and functions. Figure 10–3 illustrates a simplified view of these domain matrices. This shows five matrices at the top level

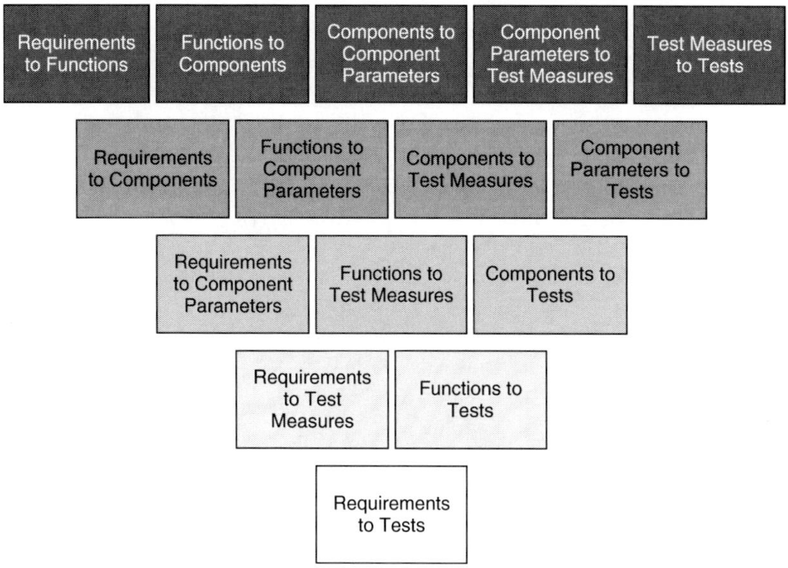

Figure 10–3 Example: design domain matrices and their resulting combinatorial matrices.

that capture six of the design domains ("working principles" as a domain is still being integrated into the design tool). The first matrix is the "requirement to function" matrix in which teams of designers list the requirements for the system of investigation and the idealized functions associated with that system. Each function is then examined with respect to each requirement to see if they are related. A binary relationship is used as opposed to other types of relationships, such as the typical 1-3-9 numbering scheme used in tools such as Quality Function Deployment (Cohen, 1995). Designers are required to only determine if a relationship exists, not to what degree it does. In this manner, the subjectivity of making qualitative decisions with low levels of uncertain information is minimized. Limited experimentation was done to support this decision in simplified academic models in-house with AID researchers (Ezhilan, 2007).

While the motivation for the development of this tool came from industry, it is important to note that exercising the tool in the evolutionary stages of tool development has been limited to interaction only with a single industry-university liaison assigned to facilitate and coordinate student design and academic research projects. In other words, this customer from industry is not a typical design engineer, but is knowledgeable about the design process used in this company and with the vehicle systems that are used to explore the tool in development. Thus, AID has relied on internal experimentation and pilot user studies to explore the limits of this design tool.

The development of this design enabler through an industry technology pull paradigm has illuminated challenges to the tool generation process. First, the need for this design tool was once again not well understood by the industry sponsor. The vision for how a requirements modeling tool can impact the design process, either by reducing design time and effort or by improving the quality or the number of concepts explored, was, and still is, entirely a "soft" vision, in that the end-users of the tool have not been consulted as AID researchers have no direct access to them. Second, where the use of such a tool is not well understood, the existing activities that will be affected by this tool also fluctuate through the internal refinement of the design process of the automotive OEM. A final challenge is that the systems chosen for evaluating the design tool are specified by the industry sponsor and not by the AID researchers. This limits the conclusions that can be drawn from the problem-focused case studies that can inform the further development of the tool. This challenge is mitigated by careful selection of internal design cases that are of little interest to the sponsor, but help the researchers in exploring the design tool. A major advantage of this approach to tool development is that, while limited, there is some industry interaction that can help in validating the design tool based on non-novice designer expertise.

Connectivity Graphs for Reverse Engineering

In two parallel design projects for a major automotive OEM, the goal was to reduce the weight of the respective components through reverse engineering. At one stage in the projects, the designers had to analyze the OEM's and its competitors' designed products. Destruction of the components to perform a Subtract and Operate Process (SOP) (Otto and Wood, 1998) was not feasible, as sub-assemblies had permanent connections between them. A need to perform "virtual" SOP was thus determined, forming the motivation to develop a design tool later named a "connectivity graph." A connectivity graph represents the components and their interconnections, with emphasis on the quantity and type of connection. Types of connection, size of component, and weight of component are represented by line type, node size, and node color. The connections may be physical (e.g., clip fit, glue) or behavioral (e.g., heat, light). Figure 10–4 shows a connectivity graph representing the assembly structure of a headlamp module. Effects of subtraction of a

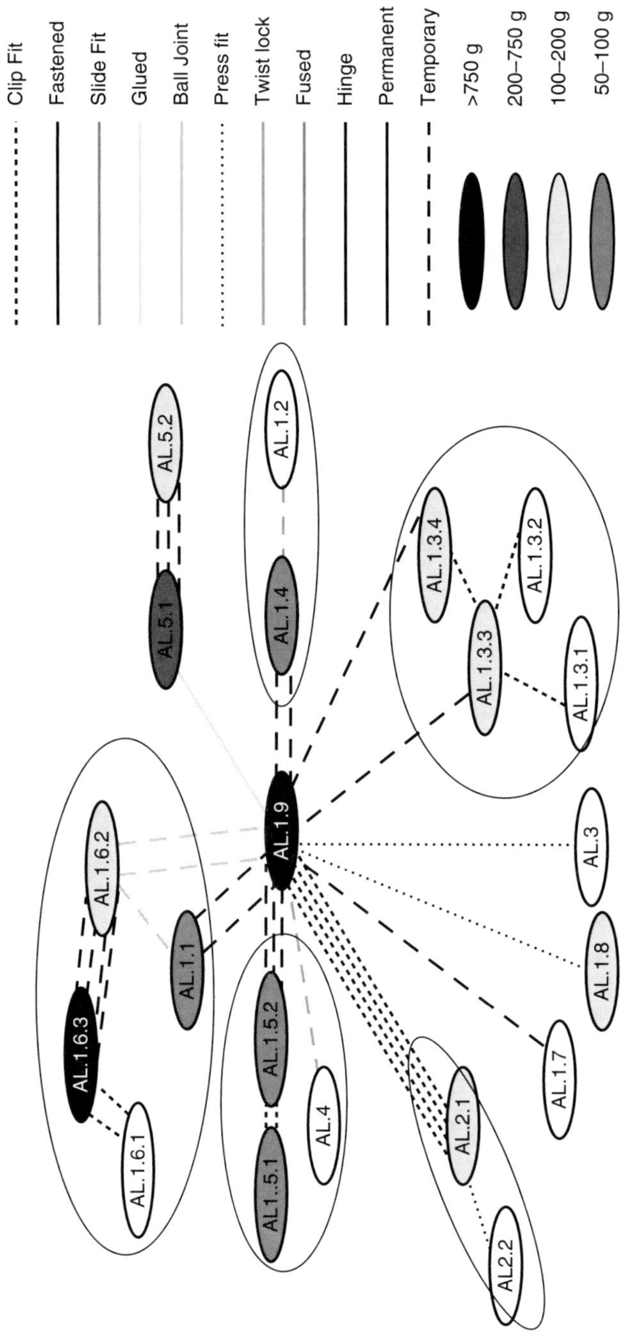

Figure 10–4 Illustration of connectivity graph for automotive headlight.

component or breaking a connection in an assembly could be studied. Redundant connections could also be found and eliminated. In some cases, as between AL.1.9 and AL.2.1, one finds that there are many connections of the same type. These multiple connections could be replaced by a single connection of a different type, making assembly and maintenance easy, thereby reducing weight. In some cases it can be seen that the connection is so critical that without it the component cannot be held in place, as with the connection between AL.1.4 and AL.1.2.

During the early stages of the design project, the connectivity graph was used to graphically represent the assembly structure of components. Algorithms to perform an effective "virtual" SOP were developed later. This limitation could have been overcome if the designers had had prior knowledge about the need of such a tool before initiating the project. As defined before, internally derived tools are project/application specific; their development resulting from experiences in design projects. Designers may have to develop special tools for different applications, which may not be reusable.

Towards an effort to validate the tool, known concepts of reducing weight were traced through the tool to examine if these concepts could be exposed by the tool (Snider, 2006). It was found that some concepts in consideration could have resulted from use of the tool. Use of the tool in other projects could provide a concrete base for validation.

The advantage of an internally derived approach to design tool development is that the motivation of these design tools is obviously known at the beginning of the development, as the design-enabler researchers are in fact the users of the design enablers. In fact, this scenario is the situation that forms the motivating vision for this research, a scenario where individual designers create their own design enablers as they go through the development process.

Validation of Design Enablers

The introduction of design enablers (DEs) may alter the design process (Crow, 2000), (Manske and Wolf, 1989). For example, solid modeling systems may implicitly direct users to create solid models in non-intuitive ways not connected with either function or manufacture. Furthermore, as the process changes, the results will change, thus improving the quality, innovation, or cost of a designed product. However, studying design is itself a complex endeavor, as illustrated in Figure 10–5, showing the user,

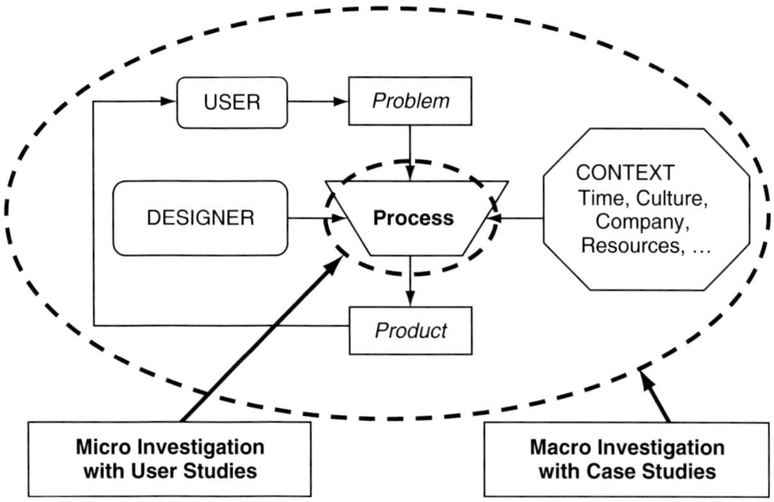

Figure 10–5 Model of design as a system with user study and case study views.

designer, problem, process, product, and context. Design enablers are created to strengthen the designer–process relationship, and it is this strengthening that must be studied. Yet the process capabilities are influenced by the design problem and context in addition to the designer and the specified process. When the entire design system is studied with case studies, the details of variables influencing design are not observable. Likewise, when user studies are employed to investigate the design process, the broader picture is lost to the researcher. Thus, each type of investigation is required in design research. Finally, while the product is the ultimate focus of industry, studying only the product cannot predict the changes to the process in relationship to context that may result from the introduction of design enablers. Therefore, to study the impact of design enablers, a set of investigative metrics is derived first and then, based on these, the impact is studied both through controlled user and industry case-studies.

Metrics

The evaluation of design enablers has primarily focused on computational metrics (e.g., time, complexity) or anecdotal evidence of improved products.

Here, however, DE metrics supplementing this nominal set are developed to predict the utility of new design enablers through quantitative and qualitative experimental validation. The derivation and justification of these metrics (monitoring, process reduction, collaborative scalability, and learning enhancement) are key components of this research, though this is not the primary focus of this paper. The metrics evaluate design enablers from the perspective of both developer and user, where, ideally, the engineer is both.

- **Monitoring:** Often, users of design enablers prefer not to relinquish total control of the process to the automated system (Nandakumar, 2000). This metric can be evaluated with industry surveys to find desired levels of transparency of design enablers, including ease of monitoring and modifying.
- **Process reduction:** Researchers can use standard techniques of non-intrusive protocol analysis to study and model the direct impact that a design enabler has on the sequence of steps, the time of execution, and the number and type of design tasks.
- **Collaborative scalability:** With the growing complexity of product design resulting from stakeholders connected across geographic and temporal boundaries, design enablers that facilitate collaboration are needed. In evaluating this metric, various issues (simultaneous users, underlying model distribution, system interoperability, representational transformation) may be used to classify design enablers.
- **Learning Enhancement:** An important potential benefit from the introduction of new DEs is that a user may develop a deeper understanding of the process, exposing DEs as potential teaching tools. A tool to help designers understand the process at the expense of efficiency has potential in an educational environment despite little to no immediate, obvious industrial return.

Experimental Verification

These general metrics combined with application-specific metrics can be systematically studied through controlled user studies. Subjects can be undergraduate and graduate engineering students, as the experiments could be integrated into courses. The AID Group has demonstrated that this is a viable approach as reflected in previous investigations on idea generation and collaborative design (Shah et al., 2001), (Ostergaard, Wetmore, and Summers, 2003), (Wetmore and Summers, 2004). As user studies can quickly become intensive, pilot studies should be employed to determine (1) if the metrics have significant relevance and (2) if they distinguish between design enablers. Following the pilot studies, larger directed experiments may then be conducted, such as studying the effectiveness of changing the level of transparency in the system. The requirements-modeling tool development included small-scale pilot studies with graduate student participation; larger-scale implementation is pending on this project.

Case Study Validation

To triangulate the quantitative results from experimentation, industrial case studies should be used to collect qualitative data on how the design enablers change the design process with respect to the identified metrics. For these studies, local companies interested in introducing new design enablers in their development processes may be recruited. This case study validation is applicable primarily with push/pull development styles. To determine the sustained impact, these case studies, through process modeling, surveys, and interviews, can capture three stages (Figure 10–6): (1) design without design enablers, (2) design with design enablers immediately after introduction, and (3) design with design enablers after assimilation and acceptance. The first stage in the case study is used to create a baseline model of the current design process and activities into which the new design enabler will be introduced. During the second stage, once the tools have been integrated into the design process, researchers should examine the initial changes to process and product. This case study can provide insights into how the users of the tool are adapting from their past approaches to work with the new tool. Challenges of usability, acceptance, and overall "buy-in" can be studied and used to inform continual tool improvement. Finally, as a longitudinal study, the companies should be revisited after enough time has passed to allow for tool acceptance in the company culture, to determine its sustained impact. Currently, only the lamelle retrieval tool project has incorporated this systematic case study research from the beginning of the project.

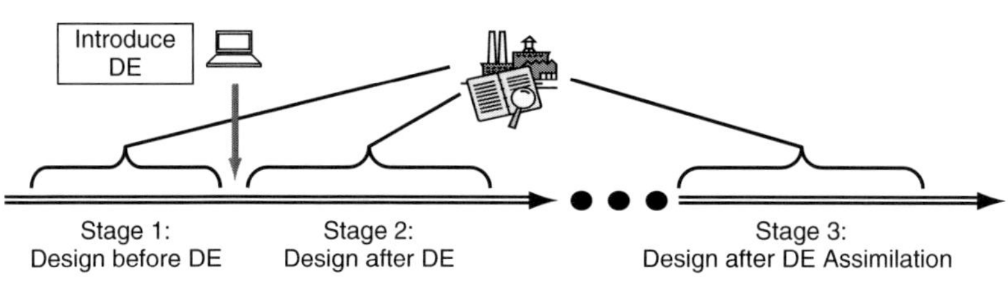

Figure 10–6 Industrial case study timeline.

TABLE 10–2 Evaluation of DEs at AID.

Design enabler	Metrics				Case study evaluation	Experimental verification
	Monitoring	Process reduction	Collaborative scalability	Learning enhancement		
Lamelle retrieval system	Low	In progress	Fair	Good	Future work	In process on a major tire company project
Requirements modeling tool	Fair	Future work	Good	Very good	Finished two case studies on automotive components	Future work
Connectivity graph	Good	Future work	Fair	Fair	Future work	Finished verification on reverse engineering projects

Evaluation of Design Enablers at AID

Table 10–2 is a summary of evaluation procedures carried out on the three design enablers that were discussed here. It can be seen that some of the design enablers fare well against certain criteria, while others do not. The criterion "process reduction" is a difficult factor to measure, as the process before and after implementation of the design enabler in consideration has not yet been completed. This type of evaluation of a design enabler may span over years. As all three of these DEs were developed recently, and thus evaluation of this criterion would take more time, they are not evaluated on this metric and are considered part of future work.

Metrics that measure characteristics of design enablers that promote innovation can also be incorporated along with the ones listed. Such an approach of evaluating design enablers during their development process facilitates modifying them accordingly, based on the feedback of their performance with respect to individual criteria.

Lessons Learned

In the course of developing these design enablers and others, we learned certain lessons that may prove beneficial in future development. These lessons include insights in the documentation of the tool development projects, how to conduct non-industry case studies, how to conduct pilot user studies using graduate students, the significance of motivation for the success of tool development, and the challenges in fitting anticipatory technologies to design enabler development.

Documentation of the design process and knowledge proved to be essential in development of any DE. The "lamelle retrieval project" transitioned from one graduate student to two different students during its two years of development. As the knowledge, methods, and process followed by the initial student were documented, the transition took place smoothly, without much time having to be spent on understanding the problem and status of the project. Major rework can be avoided by proper documentation of the development process. Documentation also helps one revisit and modify any mistakes based upon feedback and discoveries at a later stage.

User studies are employed during early phases of validation and parameter-tuning of a DE. User studies require large amounts of time and effort, as they are rigid and need to be executed following certain rules over a large sample.

Though the number of variables that can be analyzed is limited, the control on these variables is high. As an academic institution, we have a large population of undergraduate student engineers available to perform user studies. Students who practice design, however, mostly graduate students, are limited in number. Therefore, the user studies are typically focused on novice designers (undergraduates) for parameter-tuning and focused on graduate students for general validation and concept-testing. Case studies are done at a later stage of validation, usually after modifying the DE based on results from user studies, when the designer needs an in-depth, problem-specific analysis of a single implementation. Case studies performed in industry-sponsored projects are constrained by time and cost. Though they might be useful for the designer to analyze and interpret the results, it becomes difficult to report the results as it is proprietary to the company. Case studies are quick, easy to implement, easy to analyze, and flexible (Teegavarapu, Summers, and Mocko, 2008a), (Teegavarapu, Summers, and Mocko, 2008b).

Among the three DEs mentioned in this paper, the "requirements modeling tool" is the one most successful in terms of industry acceptance and use. This particular DE was built in close collaboration with industry, according to their needs, and so is seen by the customer as an added value to their design process. The connectivity graph, an extra tool (not required by the customer) developed while working on an industry-sponsored project is not used in the industry, though it has been used in various undergraduate courses to support reverse engineering and in other internal design projects in the AID Group.

Benefits from industry interaction are maximized when the customer needs to develop a DE for an existing problem, rather than wants a DE for a problem that does not yet exist, but is speculative. The "lamelle retrieval" project taught us to deal with the challenge of fitting an anticipatory DE to real-world problems. Many of the advantages and shortcomings of using the design exemplar for developing a DE came to light during this exercise.

Conclusion

Validation is a critical step in the development of a robust design enabler that is repeatable. In this paper, we attempted to emphasize various techniques followed by AID designers, their advantage and disadvantages, and lessons learned. Techniques that are suitable to a specific type of design enabler development should be chosen carefully. We urge the design community to report not only the design enablers in their final form, but the process they adopt and their findings during the course of development, in their respective publications. These findings of other design enabler developers could prove valuable for future development efforts ultimately aiming to design innovative products.

References

Altshuller, G. (1999). *The innovation algorithm*, 1st ed. Worcester, Mass.: Technical Innovation Center, Inc.

Anandan, S., Srirangam, M., and Summers, J. D. (2008). A case study in the use of the design exemplar as a search and retrieval tool. ASME Design Engineering Technical Conferences and Computers in Engineering Conferences, Brooklyn, N.Y., p. #49975.

Chavali, S., Sen, C., Mocko, G., and Summers, J. D. (2008). Using rule-based design in engineer-to-order industry: An SME case study. *Computer-Aided Design and Applications*, vol. 5, pp. 178–193.

Cohen, L. (1995). *Quality function deployment—How to make QFD work for you*, 1st ed. Boston, Mass.: Addison-Wesley.

Crow, K. (2000). *Design automation requirements to support integrated product development*, vol. 2005: NPD-Solutions, Palos Verdes, Calif.

Divekar, A. (2004). The design exemplar: The foundation for a CAD query language. In *Mechanical Engineering*. Clemson, S.C.: Clemson University.

Divekar, A., and Summers, J. D. (2003). Investigation of the design exemplar as a CAD query language. International Conference on Engineering Design, p. 1109. Stockholm, Sweden.

Ezhilan, T. (2007). Modeling requirements propagation to generate solutions for minimizing mass. *Mechanical Engineering*, MS thesis, Clemson, S.C.: Clemson University, p. 233.

Kayyar, M., Summers, J. D., Ameri, F., and Biggers, S. (2007). A case study of SME design process and development of a design enabling tool. ASME Design Engineering Technical Conferences and Computers in Engineering Conferences, Las Vegas, Nev., p. #35610.

Kim, H., Kokkolaras, M., Louca, L., Delagrammatikas, G., Michelena, N., Filipi, Z., Papalambros, P., and Assanis, D. (2002). Target cascading in vehicle redesign: A Class VI truck study. *International Journal of Vehicle Design*, vol. 29, pp. 199–225.

Maier, J., Ezhilan, T., Fadel, G. M., Summers, J. D., and Mocko, G. (2007). A hierarchical requirements modeling scheme to support engineering innovation. International Conference for Engineering Design, Paris, France.

Manske, F., and Wolf, H. (1989). Design work in change: Social conditions and results of CAD use in mechanical engineering. *IEEE Transactions on Engineering Management*, vol. 36, pp. 282–292.

Mocko, G., Summers, J. D., Teegavarapu, S., Ezhilan, T., Maier, J., and Fadel, G. M. (2007). A modeling scheme for capturing and analyzing conceptual design information: An application to the hair dryer example and comparison to existing literature. International Conference for Engineering Design, Paris, France.

NANDAKUMAR, S. (2000). Classification, parameterization, and recognition of NC features with sculptured surfaces. Mechanical and Aerospace Engineering Department. Tempe, Ariz.: Arizona State University.

OSTERGAARD, K. J. (2002). Investigation of Resistance to Information Flow in the Collaborative Design Process. *Mechanical Engineering*. Clemson, S.C.: Clemson University.

OSTERGAARD, K. J., and SUMMERS, J. D. (2007). Resistance-based modeling of collaborative design. *Concurrent Engineering Research and Applications*, vol. 15, pp. 21–32.

OSTERGAARD, K. J., WETMORE, W., and SUMMERS, J. D. (2003). A methodology for the study of the effects of communication styles on design review effectiveness. Design Engineering Technical Conferences, Chicago, Ill., p. DAC–48742.

OTTO, K., and WOOD, K. (1998). Product evolution: A reverse engineering and redesign methodology. *Research in Engineering Design*, vol. 10, pp. 226–243, December.

OTTO, K., and WOOD, K. (2001). *Product design—techniques in reverse engineering and new product design*, 1st ed. Upper Saddle River, N.J.: Prentice Hall.

PEHLIVAN, S. (2006). Representation for integration of computer aided fixture design systems. In *Mechanical Engineering*, Ph. D. dissertation, Clemson, S.C.: Clemson University.

PEHLIVAN, S., and SUMMERS, J. D. (2008). A review of computer-aided fixture design with respect to information support requirements. *International Journal of Production Research*, vol. 46, pp. 929–947.

PEHLIVAN, S., DEMPSEY, C., and SUMMERS, J. D. (2005). Information modeling for fixture design. In ASME Design Engineering Technical Conferences and Computers in Engineering Conferences, Long Beach, Calif., Paper Number: 85417.

PEHLIVAN, S., SUMMERS, J. D., and HUANG, Y. (2004). Representation requirements for supporting intelligent fixture design retrieval and reuse. Intelligent Systems in Design and Manufacturing V—Optics East 2004, Philadelphia, Pa.

PUTTI, S. (2007). Dynamic networking of design exemplars: Towards a mechanical design visual programming language. In *Mechanical Engineering*, MS thesis, Clemson, S.C.: Clemson University, p. 91.

PUTTI, S., and SUMMERS, J. D. (2006). Dynamic exemplar networks: Towards a mechanical design visual programming language. ASME Design Engineering Technical Conferences and Computers in Engineering Conferences, Philadelphia, Pa., p. #99669.

SCHUNN, C. D., PAULUS, P. B., CAGAN, J., and WOOD, K. (2006). *Final report from the NSF innovation and discovery workshop: The scientific basis of individual and team innovation and discovery*. National Science Foundation, Washington, D.C.

SHAH, J. J., FINGER, S., LU, S., LEIFER, L., CRUZ-NEIRA, C., WRIGHT, P. K., CAGAN, J., and VANDENBRANDE, J. (2004). ED2030: *Strategic plan for engineering design*. Design Automation Lab—Arizona State University, Gold Canyon, Ariz., Report March 26–29.

SHAH, J. J., VARGAS-HERNANDEZ, N., SUMMERS, J. D., and KULKARNI, S. (2001). Collaborative sketching (C-Sketch): An idea-generation technique for engineering design. *Journal of Creative Behavior*, vol. 35, pp. 168–198.

SMITH, G., TROY, T., and SUMMERS, J. D. (2006). Concept exploration through morphological charts: An experimental study. ASME Design Engineering Technical Conferences and Computers in Engineering Conferences, Philadelphia, Pa., p. #99659.

SNIDER, M. (2006). Extended toolset for reverse engineering to support lightweight engineering. *Mechanical Engineering*, MS thesis, Clemson, S.C.: Clemson University, p. 114.

SNIDER, M., and SUMMERS, J. D. (2006). Database support for reverse engineering product teardown and redesign. In CAD'06, Phuket, Thailand.

SNIDER, M., TEEGAVARAPU, S., HESSER, D., and SUMMERS, J. D. (2006). Augmenting tools for reverse engineering methods. In ASME Design Engineering Technical Conferences and Computers in Engineering Conferences, Philadelphia, Pa., p. #99676.

SUMMERS, J. D., DIVEKAR, A., and ANANDAN, S. (2006). Towards establishing the design exemplar as a CAD query language. *Computer-Aided Design and Applications*, vol. 3, pp. 523–532.

SUMMERS, J. D., SHAH, J. J., and BETTIG, B. (2004). Design exemplar: Domain independent representation structure for embodiment design automation. *Journal of Mechanical Design*, vol. 126, September, pp. 775–787.

TEEGAVARAPU, S., SNIDER, M., SUMMERS, J. D., THOMPSON, L., and GRUJICIC, M. (2007). A driver for selection of functionally inequivalent concepts at varying levels of abstraction. *Journal of Design Research*, vol. 6, pp. 218–238.

TEEGAVARAPU, S., SUMMERS, J. D., and MOCKO, G. (2008a). Case study methods for design research. ASME Design Engineering Technical Conferences and Computers in Engineering Conferences, Brooklyn, N.Y., p. #49980.

TEEGAVARAPU, S., SUMMERS, J. D., and MOCKO, G. (2008b). Establishing the need for a structured approach to developing design methods. Tools and Methods for Competitive Engineering Conference, Izmir, Turkey.

ULRICH, K. and EPPINGER, S. D. (2004). *Product design and development*, 3rd ed. New York: McGraw-Hill.

WETMORE, W. (2004). PRSM, Proper Review Selection Matrix. *Mechanical Engineering*. MS thesis, Clemson, S.C.: Clemson University, p. 179.

WETMORE, W., and SUMMERS, J. D. (2004). Influence of group cohesion and information sharing on effectiveness of design review. Design Engineering Technical Conferences, Salt Lake City, Utah, p. DAC–57509.

WETMORE, W., SUMMERS, J. D., and GREENSTEIN, J. (2008). Experimental study of influence of group familiarity and information sharing on design review effectiveness. *Journal of Engineering Design*.

ZWICKY, F. (1969). *Discovery, invention, research—through the morphological approach*. Toronto: Macmillan.

CHAPTER 11

SUPPORTING INNOVATIVE CONSTRUCTION OF EXPLANATORY SCIENTIFIC MODELS

WILL BRIDEWELL

STUART R. BORRETT

PAT LANGLEY

Consider the following scenario with characteristics common to science. An ecologist is studying an aquatic ecosystem to learn how it functions. Data gathering has yielded weekly measurements for several variables, such as the concentrations of nitrogen, phosphorus, and phytoplankton. Daily measurements exist for water temperature, solar irradiance, wind speed, and wind direction. Finally, weekly reports of zooplankton abundance exist for the summer months. Hopefully, this information will lead to a mathematical model that accurately predicts the ecosystem's response to environmental management.

Few would deny that such model-construction tasks involve creativity. The scientist must assemble a new artifact that explains the observations in a consistent and coherent way. The space of possible models is quite large, making it impractical to simply consider each candidate in turn. The problem may not be as challenging as discovering the general theory of relativity, since it does not involve paradigm shifts, but even model creation within an established theoretical framework can stretch the cognitive abilities of experienced scientists.

Luckily, the situation is not hopeless. In addition to the observed data, the ecologist also has knowledge regarding mechanisms that might plausibly operate within an aquatic ecosystem. For example, the zooplankton likely eats the phytoplankton, but the rate of consumption, the regulating factors, and the overall effects of this grazing process are undetermined. The scientist can also use deeper theoretical knowledge to guide the construction of the final model. This knowledge can consist of reasonable bounds on rates, plausible causal links, and possible formulations of grazing, amongst other things. In many cases, the ecologist will even have an existing mathematical model (e.g., Moore et al., 2002; Benz et al., 2001) that is adaptable to the current ecosystem.

Nevertheless, this remains a challenging task that could benefit from computational assistance. Current approaches to ecosystem modeling range in scope from writing custom FORTRAN programs (Arrigo et al., 2003) to using graphical model-building tools such as STELLA (Richmond et al., 1987; Sage et al., 2003). These solutions vary in difficulty of use, but the end product for each is a simulation model that one can represent as a system of differential equations. There are two primary disadvantages to using such software. The first is that one must make simultaneous decisions about which biological processes to model and how to represent them. This aspect mixes theoretical knowledge about how ecosystems operate with problem-specific assumptions relevant only in a working context. As a result, the models' complexity increases while their comprehensibility decreases. The second disadvantage is that one must build each model by hand. This requirement creates undue conservatism by contributing to a general reluctance to explore and evaluate alternative models, which in turn decreases the chances of finding innovative solutions.

We believe that concepts and methods from artificial intelligence and cognitive science suggest a better approach to designing computational aids for scientific model creation. In the pages that follow, we describe PROMETHEUS, an interactive environment for constructing and revising process accounts of dynamic systems (Bridewell et al., 2006). To clarify the rationale behind the program's design, we must recount the challenges that the problem presents to intelligent assistants.

First, we should note that despite recent rhetoric in the data-mining literature scientific data relevant to discovery are often rare and difficult to obtain. The costs of collecting and preparing the data are non-trivial, and high rates or long periods of sampling may be impossible. As a result, the number of samples probably ranges in the low hundreds. Given the number of variables, parameters, and relationships in the target models, common methods for data mining are inappropriate, and we require new techniques.

Another challenge requires us to support model-revision in terms of both causal structure and parameters. Systems scientists like our ecologist come to a modeling task with prior knowledge of various sorts. At one level, this knowledge consists of the possible interactions among entities in a system and ways to formulate those relationships. For example, the ecologist knows that a process of phytoplankton growth exists and that it must be included in the final model. However, whether this growth can best be modeled as exponential, logistic, or something more complex may be unknown. At a different level, the ecologist may seed the discovery process with a prior model and search for revisions that explain the current data.

The third challenge is the need for *communicable* models. As we mentioned, ecologists often express their models in terms of differential and algebraic equations, but machine learning traditionally uses its own notations (e.g., decision trees, logical rules, Bayesian networks), which result in models that are not easily communicated to domain scientists. We need techniques for knowledge discovery that produce output that closely approximates the scientists' own modeling language.

In addition, scientists want models that move beyond description to provide *explanations* of their data. Regression-style techniques generate pithy summaries of the observations, but they fail to make contact with the underlying generating mechanisms. This desire poses the challenge of developing methods that construct explanatory models rather than purely descriptive ones.

These issues raise algorithmic challenges, but introspection suggests another problem. Many computational discovery systems strive to automate the activity of model construction, but few scientists want to be replaced. However, they may well accept computational tools that carry out tedious aspects of searching through the model space, provided we find ways they can participate in the model-building endeavor. Ideally the software would perform lower-level tasks and free the scientist to concentrate on higher-level goals. Thus, it behooves us to design interactive systems that support a creative partnership between software and scientific domain experts.

This chapter describes the application of ideas from artificial intelligence and cognitive science approaches to stimulate discovery in the systems sciences like ecology. As such, it introduces the above challenges and our response as embodied in PROMETHEUS, an environment that supports the creation of quantitative models of dynamic systems. The next section describes challenges in user interaction and our responses. We then discuss the challenges in developing a model discovery system, highlighting the integration of various threads of research to compose an intelligent assistant for scientific modeling. After this, we briefly discuss previous results from the use of PROMETHEUS and identify new challenges that have arisen during experimentation. Finally, we summarize our work and highlight unmet challenges that seem ripe for further research.

ADDRESSING CHALLENGES IN COMMUNICATION

One should address challenges of user interaction from the foundation upwards when building an intelligent system. *To meet the challenges of model comprehensibility and explanation,* PROMETHEUS *represents its knowledge in a language that builds on systems of equations.* Models expressed as differential and algebraic equations commonly appear in the ecosystems literature and pervade systems science as a whole. However, even in this familiar form, the explanatory content of the models is not easily accessible. Fortunately, we can turn to ecology for a solution. The models in this domain often portray mechanisms (e.g., Gaff et al., 2004; Sarmiento et al., 1998), which suggests that the language of entities and the processes in which they participate (Machamer et al., 2000) is appropriate. Forbus (1984) previously developed a formalism for qualitative process models, which takes this basic perspective, but our purposes, which include close contact with numeric data, suggest a need for *quantitative* process models.

Representing the models as mechanisms also addresses the challenge of a participatory system. Although systems of equations are the output of this task, scientists initially work at a conceptual level. For instance, Jørgensen and Bendoricchio (2001) recommend developing a conceptual structure of the studied system as the first step in ecological modeling. They suggest building this structure by listing the state-variables and then identifying the physical, chemical, and biological processes that link the variables to each other and to the environment. Afterwards, one uses mathematical formulations of the processes to produce an equivalent system of equations. We want to support this modeling style that gives scientists the creative freedom to design the larger-scale features of the modeled system before making low-level decisions about the nature of the processes.

Finally, the quantitative process representation also addresses a technical challenge. Unlike previous modeling environments, PROMETHEUS supports automated search through the space of models. The space of differential equations is far too large for unguided search, and it most certainly contains models that fit the observed data but lack plausibility. The processes used by PROMETHEUS contain meaningfully grouped chunks of equations that one can combine with others to form the model. For instance, a process describing predation between species would have one equation element that decreases the prey population and another that increases the predator population. Therefore, removing such a process would completely excise predation from the model and update the system of equations appropriately. By defining these processes, one can use knowledge from systems science to restrict PROMETHEUS's search to a space of plausible models.

Both the entities and the processes in quantitative process models have two forms: generic and instantiated. A generic entity, as shown in Table 11–1, declares the variables and parameters that store relevant properties. Parameters at both the process and entity levels are immutable, model-specific values that fall within a specified range. In contrast, the variable values can change over time. Variables themselves fall into one of three classes. An *exogenous* variable can only influence processes in the model, and its values must be read from a data source. An *observed* variable must be

TABLE 11–1 The generic entity for a primary producer contains a measure of its species' concentration, growth rate, and loss rate. Processes affecting the concentration will have additive influence, whereas the current growth rate will be the minimum of values produced by multiple processes. The loss rate must fall between zero and ten.

```
generic entity primary_producer:
  variables:
    conc {sum}
    growth_rate {min}
  parameters:
    loss_rate [0, 10]
```

explained by the model and must have associated data for purposes of comparison, and an *unobserved* variable needs only an initial value and a range in which this value should fall. All variables and parameters associated with an entity are passed along with that entity to any process in which it participates. One can instantiate a generic entity by specifying whether each variable is observed, unobserved, or exogenous; identifying necessary data sources; and assigning a numerical value to each parameter.

Generic processes contain entity and process roles, parameters, conditions, and equations. Entity roles consist of a local name for an entity along with the number and types of entities that can fill that role. For instance, the exponential loss process in Table 11–2 requires a single generic entity that has type "primary producer" or "grazer." A process role gives a process type and the list of entities to pass along to the selected subprocess. In addition, Boolean conditions control whether a process is active based on the current value of variables in the model, and equation elements define the quantitative behavior of the process. As a final feature, each generic process has a type that helps guide the search for plausible subprocesses. The instantiated form of a process requires one to specify the participating entities, any subprocesses, and local parameter values.

Generic processes and entities address the challenges of incorporating prior knowledge and model discovery with few data. The generic components along with the constraints among them limit the model space to a subset of plausible structures, and this tight restriction helps offset the difficulties of knowledge discovery from small data sets. The structural constraints manifest in three ways. First, the use of generic entities along with entity roles constrains the viable participants in a

TABLE 11–2 The generic process for exponential loss has type "loss" and takes exactly one entity with type primary producer or grazer. The single equation in this process states that the first derivative of the concentration with respect to time is equal to a loss influenced by the species' loss rate.

```
generic process exponential_loss {loss}:
  entity_roles:
    S {primary_producer, grazer} <1 to 1>
  equations:
    d[S.conc, t, 1] = −1 * S.loss_rate * S.conc
```

process. Second, the bounds on parameter values help guide estimation tools, which we will discuss in the next section. Third, the hierarchy imposed by process types and subprocesses defines a modified AND/OR tree of possible structures. The subprocesses, which may be optional, specify the AND branching and specify which process types must occur along with the current generic process. These process types establish exclusive OR branches, specifying a set of generic processes that may satisfy a particular process role. To illustrate, the process type "growth" may have several forms (e.g., exponential, logistic, limited). In this case, suppose that a top-level process called "ecosystem" requires a growth process. This need constitutes an AND branch of the tree, whereas the multiple processes of the correct type compose the OR branch.

The creation of quantitative process models requires multiple steps. Initially, a scientist must develop a library of generic processes and entities. In our experience, one begins this task at an abstract level by identifying the entities and processes relevant to a chosen context (e.g., aquatic ecosystems). Next, one specifies the mathematical forms of the processes, selects the important properties of the entities, and determines the structural constraints for the hierarchy. Much of this work is straightforward. For instance, the process forms appear in the literature, and the generic entities relate directly to theoretical terms and the measurements one would typically make in the domain. However, the constraints encoded in the process hierarchy reflect implicit knowledge and are more difficult to elicit. In addition, the syntax of the constraint-specification language can influence the organization of equations into processes and properties into entities. As a result, assembling a library involves an iterative refinement of one's knowledge and increases in difficulty with the complexity of the process hierarchy. Fortunately, once completed, a single library describes the theoretical knowledge for a sizable range of problems. Therefore, one can build multiple models from a single library, make minor adaptations to fit similar domains, and borrow components for use in other problems.

To create a model from a domain-specific library, one selects the relevant entities to instantiate, the processes that link these entities, and the particular process alternatives that drive the observed dynamics. This step may constitute a stopping point, but it is more likely that the scientist will compare the model to some observations and adjust the model as necessary. We are developing PROMETHEUS to support as much of this procedure as possible.

Addressing Challenges in Learning

PROMETHEUS consists of two major components—the user interface and the model-induction engine—each with its own set of challenges. Here we mainly discuss our approach to model construction and revision, but we first describe PROMETHEUS's wide range of interaction. At a basic level, one can create a model, view its causal flow as shown in Figure 11–1, and see the current system of equations. Additionally, the program supports model evaluation through the inclusion of a simulation engine and a means to compare the resulting trajectories with observed data. Moreover, one can manually revise models by altering parameters and adding or deleting both processes and entities. Thus, at its core, PROMETHEUS supports creativity through the ability to freely design and test quantitative process models.

At this level, PROMETHEUS operates much like other modeling packages (apart from its emphasis on mechanisms), but the integration of system identification and artificial intelligence components set it apart. These elements provide support for automated parameter estimation, model construction, and model revision. Todorovski et al. (2005) describe the underlying algorithm for these features.[1] This approach operates in two separable stages, the first of which defines the symbolic space of model structures. Beginning with the root process, PROMETHEUS satisfies the minimal set of constraints imposed by the hierarchy by including all required processes and no optional ones. This step produces a set of model structures that relate entities and processes but lack values for the parameters. At this level of the search, we predominantly draw on traditional, symbolic techniques from artificial intelligence. Specifically, the program performs a beam search through the AND/OR space defined by the background knowledge and guided by a quantitative measure of fit (i.e., sum of squared error or variance normalized mean-squared error).

For each structure, PROMETHEUS searches a second space defined by the numeric parameters. We use techniques from system identification to perform a gradient-descent search based on the quantitative measure of fit. The core algorithm, which was designed by Bunch et al. (1993), fits

[1] PROMETHEUS's current interface uses an earlier induction algorithm that lacks support for entities and process hierarchies. We are adapting the environment to use these structures.

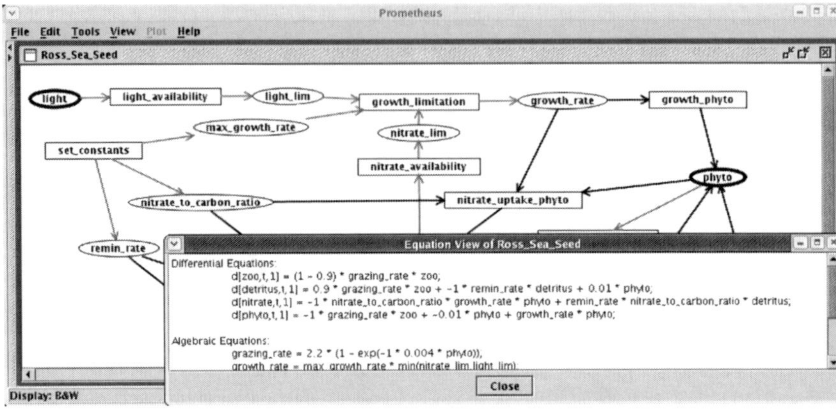

Figure 11–1 PROMETHEUS can display both a causal diagram of a model and the underlying equations. In the diagram, the ovals are variables and the rectangles are processes.

the parameters of dynamic, nonlinear systems of equations while ensuring that the resulting values fall within specified bounds. This algorithm performs a local search, so the system lets one specify a number of restarts that each explore the parameter space from a randomly selected point. In practice, we have found this approach to run slowly and to have high variance, which influences the selection of model structures. The FUSE algorithm (Bridewell et al., 2005) integrates research on ensemble methods to reduce overall variance, but we have yet to incorporate this solution into PROMETHEUS.

PROMETHEUS *meets the challenge of model revision by providing the scientist with several controls to influence semi-automated revision.* As input, the scientist provides an initial model along with three lists: (1) processes that may be removed, (2) generic processes that may be instantiated, and (3) processes and entities whose parameters may be changed. The structural search uses the initial model with all deletable processes removed to seed the search. From that point on, the algorithm tries both to add deleted processes back to the model and to add instantiations of the specified generic processes when possible. For the most part, revision operates just like induction from scratch, but the scientist's guidance further limits the possible moves in the search space. Upon completion, the program returns a list of the best models ranked by the chosen measure of quantitative fit. Each of these models can serve as a foundation for future revisions.

We can best describe the use of PROMETHEUS by example. Consider the ecologist described at the beginning of this chapter. This modeler begins by identifying a set of generic entities and processes expected to operate within the observed ecosystem. One could draw this knowledge from an earlier developed library, extract it from textbooks or articles, or create it a new. After developing this library, the ecologist can build an initial model in PROMETHEUS. The model may contain nothing more than a list of the entities, or it could be fully detailed, with all suspected relationships indicated with instantiated processes. For this example, we will assume the second case.

With a model structure in place, the ecologist can then fit the parameters using all available data and simulate the resulting model to compare the output with observations. Now, suppose that the scientist notices that the simulated phytoplankton population fails to decrease as expected. Examination of the model shows that nothing grazes on the phytoplankton, even though zooplankton exist in the region under study. The ecologist can either manually select and add the grazing process or have PROMETHEUS search the reduced space of models consisting of the initial structure plus all possible options for the inclusion of grazing. If the user opts for automated revision, the program will yield a ranked list of plausible models. The scientist may select, simulate, and evaluate each of the results, and if necessary, the revision process can continue.

Importantly, PROMETHEUS transforms the modeling task by automating lower-level tasks such as assembling equations, fitting parameters, and generating alternatives. Instead, the ecologist can concentrate on the types of processes likely to appear in an ecosystem, their alternative functional forms, and the constraints among the processes. More directly, the automated search tools in PROMETHEUS let one work closer to the theoretical structures and modeling assumptions that characterize plausible explanations. Given this information, the software explores the space of candidates, highlighting those few that both fit the background knowledge of the domain and match available observations.

INITIAL EXPERIENCES WITH PROMETHEUS

Researchers have evaluated PROMETHEUS's behavior in a variety of scientific domains. In this section, we summarize the nature of the tasks, the results obtained with the system, and some lessons suggested by those experiences. We focus on model induction in our description of two scientific tasks, and discuss an application of model revision to the Ross Sea domain. Detailed results appear in earlier papers, so here we present only the highlights.

Predator–Prey Interactions in Protists

Predator–Prey systems are among the simplest in ecology, which makes them a good starting point for evaluating PROMETHEUS. In earlier work (Asgharbeygi et al., 2006; Todorovski et al., 2005), we explored the protist system composed of the predator *Didinium nasutum* and the prey *Paramecium aurelia* using data from experiments originally reported by Veilleux (1979). Jost and Ellner (2000) report the observed values, which consist of population concentrations recorded in 12-hour intervals for three experimental conditions. The data, some of which appear in Figure 11–2, are fairly smooth and exhibit oscillatory behavior.

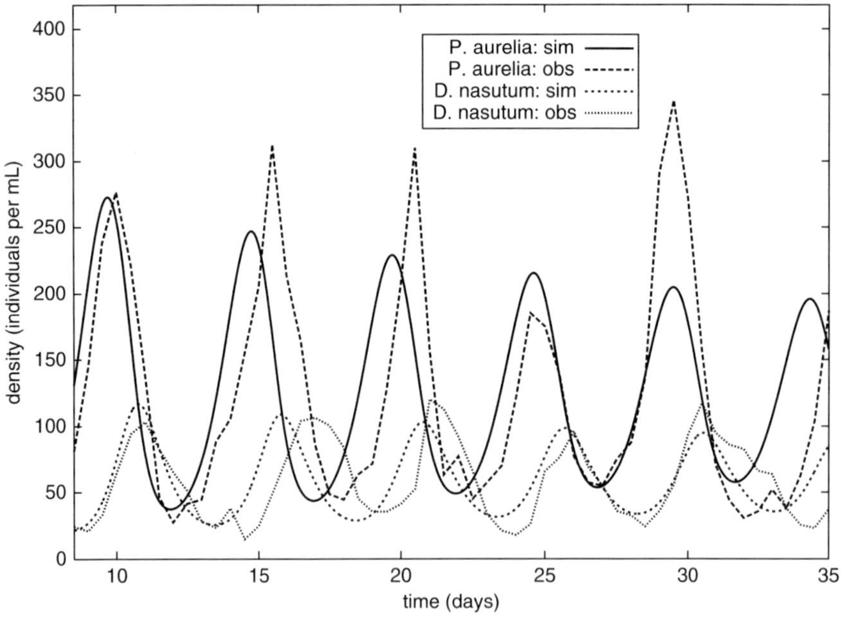

Figure 11–2 Population dynamics in a simple predator–prey ecosystem.

For this domain, we provided PROMETHEUS with generic processes for prey growth, predator decay, and predation, including alternative functional forms. When constrained by the process hierarchy, these defined a space of 24 distinct model structures that, with parameters specified, predict trajectories for the two species' concentrations from their initial values. The system's search of this space produced a plausible model that included processes for growth, predation, and decay. As shown in Figure 11–2, the simulated curves track the heights and timing of the observed trajectories reasonably well.

Notably, we encountered problems when we presented the system with the entire Jost and Ellner data set, and obtained these results only when we provided it with a selected subset. Measurements early in the time series had considerably lower peaks, which suggested a different regime was operating for unknown reasons. This result reveals an important ability that PROMETHEUS currently lacks: *When a scientific modeling system cannot explain an entire set of observations, it should consider ignoring some of the data.* This capability could help the system both identify separate regimes and minimize the effects of outliers during the early stages of modeling. Clearly, human scientists have this capacity, and future versions of PROMETHEUS would benefit from a solution that meets this challenge.

Population Dynamics in the Ross Sea

The Ross Sea in the Southern Ocean involves a somewhat more complex ecosystem. Here the phytoplankton, which may play an important role in the global carbon cycle (DiTullio et al., 2000), undergo repeated cycles of population increase and decrease. In this case (Asgharbeygi et al., 2006), we had access to two sets of 188 daily measurements for phytoplankton that spanned two successive years. Concurrent data were also available for nitrate concentrations and ice coverage; we used an algebraic equation to simulate the light dynamics.

Based on discussions with the team's biological oceanographer (Kevin Arrigo), we identified entities of interest and developed 25 generic processes that encoded how they might interact. In addition to phytoplankton and nitrate, the entities included detritus, which results from phytoplankton decay, and zooplankton, which feeds on phytoplankton. Because neither were measured, the researchers treated attributes of both as unobserved theoretical variables. In addition, they seeded PROMETHEUS with an initial model that substantially reduced the size of the structural search space.

PROMETHEUS produced a number of models that made sense ecologically and that fit the first year's data closely, but they generalized poorly to the second year's observations.

Inspection of the model suggested that ice differences across the years had little effect on phytoplankton growth, although this had originally seemed a likely explanation of differences between the two years. Discussion with the oceanographer led the group to include another generic process, which states that phytoplankton's absorption of nitrate depends on available light. Based on this information, PROMETHEUS found another model that fit the first year's data nearly as well as the earlier candidate but that, as Figure 11–3 shows, generalized much better to the second year. The implication is that the nitrogen-to-carbon ratio for phytoplankton varies as a function of light availability, which the oceanographer believes is an important ecological claim.[2] The original vision for PROMETHEUS was that it

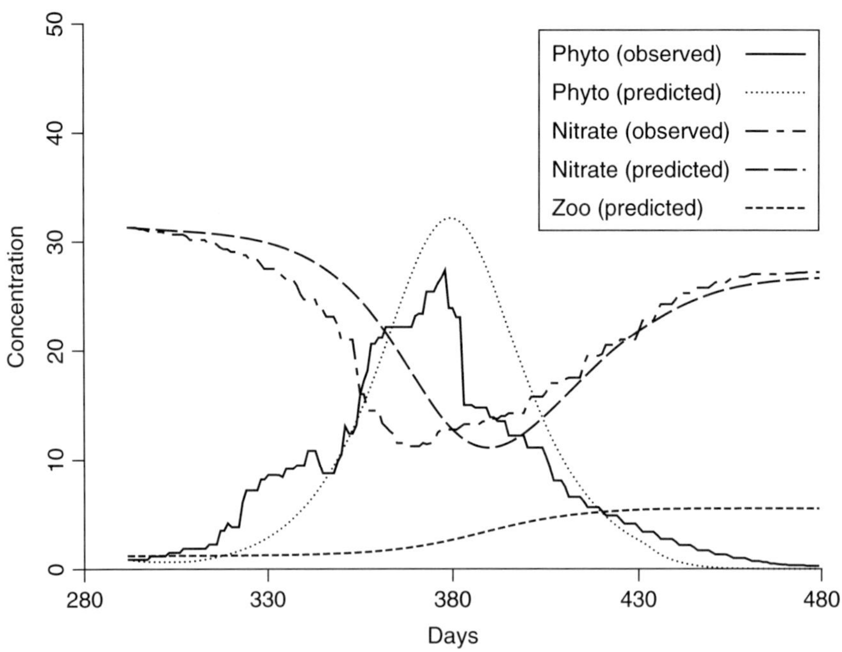

Figure 11–3 Performance on test data from the Ross Sea.

[2] This finding was made before support for entities and process hierarchies was complete.

should support the scientist's search for models in a well-defined space. However, our experience with the Ross Sea revealed another key ability that the system lacks: *When a scientific modeling system cannot account for observed differences, it should consider new mechanisms that expand its space of plausible models.* Human scientists prefer to explain phenomena in terms of familiar mechanisms, but they can consider new processes when necessary, presumably by falling back on more general knowledge. Adding such a capability to PROMETHEUS is another important direction for future work.

Biochemical Kinetics

We also applied PROMETHEUS to a problem from biochemical kinetics (Langley et al., 2006), which studies physiological changes in metabolites over time. Here we drew upon time-series data collected by Torralba et al. (2003) about the glycolysis pathway, which converts glucose into pyruvate and which plays an essential role in most life forms. Torralba's group used an impulse-response method that, given a biochemical system in steady state, briefly increases the inflow of one substance and measures its effects on others over time. We used 14 data points for six distinct glycolitic metabolites.

For this domain, we provided the system with five generic processes that encoded four types of metabolic reactions appearing in pathway models. These differ in how they affect positive and negative fluxes (i.e., flow into and out of a reaction pathway) of the substances involved. The researchers crafted four generic processes—*irreversible*, *reversible*, *inhibition*, and *activation* reactions—along with a fifth that stated a metabolite's concentration changes as a weighted sum of its positive and negative fluxes, with each flux term being multiplied by its respective rate.

When provided with the data and these generic processes, PROMETHEUS searched a space of 172 distinct models and estimated parameters for each candidate. Figure 11–4 shows both the observed trajectories and those predicted by the best-scoring model, which produces good fits in both qualitative and quantitative terms. However, the model structure differs from the generally accepted glycolysis pathway in that it lacks inhibition and activation processes. Presumably, this occurred because the system could not introduce unobserved entities to serve as inhibitors and activators, which suggests another limitation: *A scientific modeling system should consider introducing theoretical entities that augment those provided by the user.* PROMETHEUS can already generate

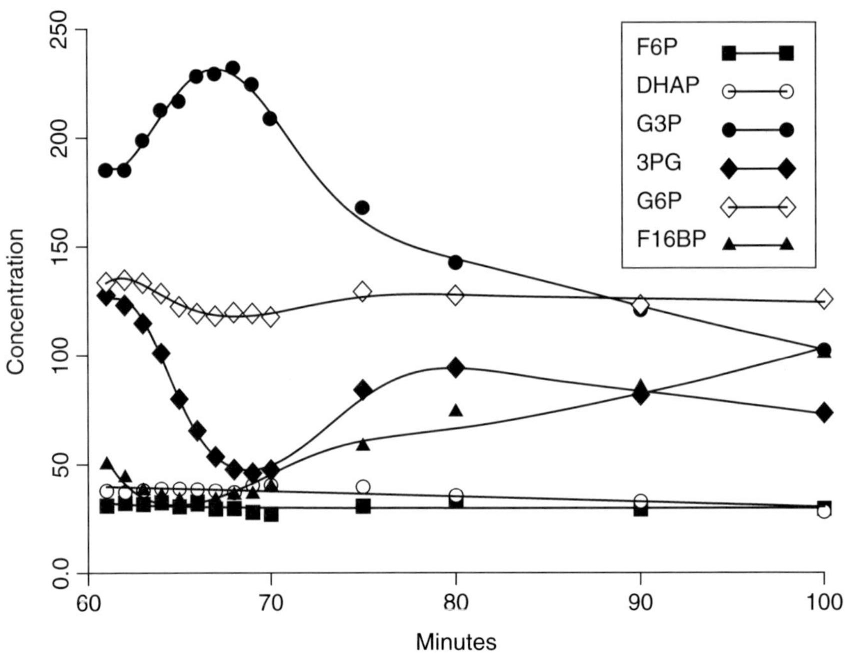

Figure 11–4 Observed (points) and predicted (lines) trajectories of chemical concentrations in the biochemical kinetics domain.

models with unobserved terms, but only when they are given as input. Introducing the ability to postulate new entities, as constrained by background knowledge, would extend the system's ability to generate plausible explanatory models.

Discussion

At the outset, we described five challenges that arise when building a tool to support the construction of scientific models. These included sparsity of relevant data, the presence of prior models and knowledge, a match between system output and the primary domain language, the production of explanatory models, and an emphasis on interactivity. We designed the formalism for quantitative process models and generic processes with these challenges in mind, and we integrated techniques from artificial intelligence and system identification in response.

The formalism for quantitative process models has some clear advantages. First, one can directly translate the models into a more familiar representation for scientists, thereby addressing the challenge of communication. Second, casting the domain knowledge as processes leads to mechanisms that explain the studied system's behavior. Finally, the processes mesh well with the conceptual stage of model-building, which eases the input of domain knowledge and prior models to the program.

To meet the challenges involved in model construction and revision, we borrowed from several research traditions. Heuristic search of AND/OR trees provides a means for navigating the space of model structures, while tools from system identification (e.g., Åström and Eykhoff, 1971) direct search through the parameter space. The use of prior knowledge helps constrain search to produce plausible models even without large data sets. Finally, theory revision techniques (e.g., Ourston and Mooney, 1990) support interactive search, letting the user gauge the scope and nature of revisions at each step in the modeling process.

Experiments with PROMETHEUS identified several open challenges for the artificial intelligence community. First, we need a way to ignore connected sets of data, not just isolated outliers, that may stem from a different regime and keep a program from producing good models. In dynamic systems, assigning observations to different operating regimes will allow easier identification of the active mechanisms. Second, a program should be able to introduce new processes to its library. Third, model construction methods should introduce theoretical entities that are not specified explicitly by the user. These last two additions can increase the search space substantially, so we need more intelligent mechanisms to guide the structural search.

Perhaps the biggest surprise we encountered involved current software capabilities. In the early stages of our work, we believed that techniques for parameter estimation were ready for application. However, we found the tools available for nonlinear dynamical systems to be both unreliable and slow. Generally, parameter estimation techniques use very little knowledge, and we believe that ideas from artificial intelligence and knowledge-based reasoning could improve these systems on both fronts. One possibility is to incorporate scientists' knowledge of both the general shape that trajectories should take and the relationships among trajectories and parameters. Bradley et al. (2001) explored another possibility that used heuristics to avoid unnecessary parameter estimation. Capitalizing on this type of knowledge is the strength of artificial intelligence, and innovations in this area will have broad applicability.

In summary, we have seen that PROMETHEUS introduces a number of innovations that respond directly to the outlined challenges and support creative acts in science. These include a representation for models and background knowledge that supports communication with scientists, integration of domain knowledge to guide symbolic and numerical search, and incorporation of initial models and user input to guide revision. However, we have also seen that this combination of ideas does not exhaust the ways that we can support the creative activities of scientists as they develop models of dynamic systems. We need additional research that extends the power and flexibility of the modeling methods to better serve the needs of scientists.

ACKNOWLEDGMENTS

This research was supported by Grant Number IIS-0326059 from the National Science Foundation. We thank Kevin Arrigo for his expertise on the Ross Sea, and Nima Asgharbeygi, Oren Shiran, and Ljupčo Todorovski for contributions to the PROMETHEUS system.

REFERENCES

ARRIGO, K. R., WORTHEN, D. L., and ROBINSON, D. H. (2003). A coupled ocean-ecosystem model of the Ross Sea: 2. Iron regulation of phytoplankton taxonomic variability and primary production. *Journal of Geophysical Research–Oceans*, 108(C7), 3231, doi: 10.1029/2001JC000856.

ASGHARBEYGI, N., BAY, S., LANGLEY, P., and ARRIGO, K. (2006). Inductive revision of quantitative process models. *Ecological Modelling*, *194*, 70–79.

ÅSTRÖM, K. J., and EYKHOFF, P. (1971). System identification—a survey. *Automatica*, *7*, 123–167.

BENZ, J., HOCH, R., and LEGOVIC, T. (2001). ECOBAS: Modelling and documentation. *Ecological Modelling*, *138*, 3–15.

BRADLEY, E., EASLEY, M., and STOLLE, R. (2001). Reasoning about nonlinear system identification. *Artificial Intelligence*, *133*, 139–188.

BRIDEWELL, W., BANI ASADI, N., LANGLEY, P., and TODOROVSKI, L. (2005). Reducing overfitting in process model induction. *Proceedings of the Twenty-Second International Conference on Machine Learning* (pp. 81–88). Bonn, Germany.

BRIDEWELL, W., SÁNCHEZ, J. N., LANGLEY, P., and BILLMAN, D. (2006). An interactive environment for the modeling and discovery of scientific knowledge. *International Journal of Human–Computer Studies*, *64*, 1099–1114.

BUNCH, D. S., GAY, D. M., and WELSCH, R. E. (1993). Algorithm 717: Subroutines for maximum likelihood and quasi-likelihood estimation of parameters in nonlinear regression models. *ACM Transactions on Mathematical Software, 19,* 109–130.

DiTullio, G. R., Grebmeier, J. M., Arrigo, K. R., Lizotte, M. P., Robinson, D. H., Leventer, A., Barry, J. B., Van Woert, M. L., and Dunbar, R. B. (2000). Rapid and early export of *Phaeocystis antarctica* blooms in the Ross Sea, Antarctica. *Nature, 404,* 595–598.

Forbus, K. (1984). Qualitative process theory. *Artificial Intelligence, 24,* 85–168.

Gaff, H., Chick, J., Trexler, J., DeAngelis, D., Gross, L., and Salinas, R. (2004). Evaluation of and insights from ALFISH: A spatially explicit, landscape-level simulation of fish populations in the Everglades. *Hydrobiologia, 520,* 73–87.

Jørgensen, S. E., and Bendoricchio, G. (2001). *Fundamentals of ecological modelling.* New York: Elsevier.

Jost, C., and Ellner, S. (2000). Testing for predator dependence in predator–prey dynamics: A non-parametric approach. *Proceedings of the Royal Society of London B, 267,* 1611–1620.

Langley, P., Shiran, O., Shrager, J., Todorovski, L., and Pohorille, A. (2006). Constructing explanatory process models from biological data and knowledge. *Artificial Intelligence in Medicine, 37,* 191–201.

Machamer, P. K., Darden, L., and Craver, C. F. (2000). Thinking about mechanisms. *Philosophy of Science, 67,* 1–25.

Mooney, R. J., and Ourston, D. (1994). A multistrategy approach to theory refinement. In R. S. Michalski and G. Teccuci (Eds.), *Machine learning: A multistrategy approach* (pp. 141–164). San Mateo, Calif.: Morgan Kaufmann.

Moore, J. K., Doney, S. C., Kleypas, J. A., Glover, D. M., and Fung, I. Y. (2002). An intermediate complexity marine ecosystem model for the global domain. *Deep-Sea Research II, 49,* 403–462.

Ourston, D., and Mooney, R. J. (1990). Changing the rules: A comprehensive approach to theory refinement. *Proceedings of the Eighth National Conference on Artificial Intelligence* (pp. 815–820). Detroit, Mich.

Richmond, B., Peterson, S., and Vescuso, P. (1987). *An academic user's guide to STELLA.* Lyme, N.H.: High Performance Systems.

Sage, R. W., Patten, B. C., and Salmon, P. A. (2003). Institutionalized model-making and ecosystem-based management of exploited resource populations: A comparison with instrument flight. *Ecological Modelling, 170,* 107–128.

Sarmiento, J. L., Hughes, T. M. C., Stouffer, R. J., and Manabe, S. (1998). Simulated response of the ocean carbon cycle to anthropogenic climate warming. *Nature, 393,* 245–249.

Todorovski, L., Bridewell, W., Shiran, O., and Langley, P. (2005). Inducing hierarchical process models in dynamic domains. *Proceedings of the Twentieth National Conference on Artificial Intelligence* (pp. 892–897). Pittsburgh, Pa.

Torralba, A. S., Yu, K., Shen, P., Oefner, P. J., and Ross, J. (2003). Experimental test of a method for determining causal connectivities of species in reactions. *Proceedings of the National Academy of Sciences, 100,* 1494–1498.

Veilleux, B. G. (1979). An analysis of predatory interaction between *Paramecium* and *Didinium. The Journal of Animal Ecology, 48,* 787–803.

Index

Note: In this Index, tables are indicated by "*t*", figures by "*f*".